THEORIES on the
SCRAP HEAP

John Losee

THEORIES on the SCRAP HEAP

Scientists and Philosophers on the
Falsification, Rejection, and
Replacement of Theories

University of Pittsburgh Press

Published by the University of Pittsburgh Press, Pittsburgh, Pa., 15260

Manufactured in the United States of America

Printed on acid-free paper

10   9   8   7   6   5   4   3   2   1

LIBRARY OF CONGRESS CATALOGING-IN-PUBLICATION DATA

Losee, John.
    Theories on the scrap heap : scientists and philosophers on the falsification,
    rejection, and replacement of theories / John Losee.
        p.   cm.
    Includes bibliographical references and index.
    ISBN 0-8229-5873-2 (acid-free paper)
        1. Science—Philosophy.   2. Science—Theory reduction.   3. Science—Research.
    I. Title.
    Q175.L666 2005
    501—dc22

Contents

Acknowledgments

Grateful acknowledgment is made of the publishers of the following works for their permission to reprint excerpts:

From Hans A. Bethe, "Energy Production in Stars," *Phys. Rev. 55* (1939), 103. Copyright © 1939 by the American Physical Society. Reprinted with permission of the American Physical Society.

From Niels Bohr, *Atomic Theory and the Description of Nature* (Cambridge: Cambridge University Press, 1961), 36–37. Reprinted with permission of Cambridge University Press.

From Peter Clark, "Atomism *versus* Thermodynamics," in *Method and Appraisal in the Physical Sciences: The Critical Background to Modern Science, 1800–1905*, ed. Colin Howson (Cambridge: Cambridge University Press, 1976), 45–46; 47–49; 59–60; 60–61. Reprinted with permission of Cambridge University Press.

From Paul Dirac, *The Principles of Quantum Mechanics* (Oxford: Clarendon Press, 1930), 1–4. Reprinted with permission of Oxford University Press.

From Paul Feyerabend, *Against Method* (London: NLB, 1975), 23–24. Reprinted by permission of the publisher.

From William Glen, "How Science Works in the Mass-Extinction Debates," in *The Mass-Extinction Debates, How Science Works in a Crisis*, ed. William Glen (Stanford, Calif.: Stanford University Press, 1994), 88–91. Copyright © 1994 by the Board of Trustees of the Leland Stanford Jr. University.

From Stephen J. Gould, "The Origin and Function of 'Bizarre' Structures: Antler Size and Skull Size in the 'Irish Elk', *Megaloceros Giganteus*," *Evolution 28* (1974), 216. Permission courtesy of the Society for the Study of Evolution.

From Joel Kingsolver and M. A. R. Koehl, "Aerodynamics, Thermoregulation, and the Evolution of Insect Wings: Differential Scaling and Evolutionary Change," *Evolution 39* (1985), 489–90. Permission courtesy of the Society for the Study of Evolution.

From Imre Lakatos, "History of Science and its Rational Reconstructions," in *Boston Studies in the Philosophy of Science,* vol. 8, *In Memory of Rudolf Carnap,* pp. 91–136, ed. R. Buck and R. S. Cohen (Dordrecht: Reidel, 1971), 100–101, 104–5, 108–22. Copyright © 1971 by D. Reidel Publishing Co., Dordrecht, Holland. Reprinted by permission from Kluwer Academic Publishers.

From Richard Leakey, *The Origin of Humankind.* (New York: Basic Books, 1994), 87; 95–97. Copyright © 1994 by B. V. Sherma. Reprinted by permission of Basic Books, a member of Perseus Book, L.L.C.

From Alan Musgrave, "Method or Madness?" in *Boston Studies in the Philosophy of Science,* vol. 39, *Essays in Memory of Imre Lakatos,* ed. R. S. Cohen et al. (Dordrecht: Reidel, 1976), 458–62; 466–67. Copyright © 1976 by D. Reidel Publishing Co., Dordrecht, Holland. Reprinted by permission from Kluwer Academic Publishers.

From Alan Musgrave, "Why Did Oxygen Supplant Phlogiston? Research Programmes in the Chemical Revolution," in *Method and Appraisal in the Physical Sciences: The Critical Background to Modern Science, 1800–1905,* ed. Colin Howson (Cambridge: Cambridge University Press, 1976), 187–91; 203. Reprinted with permission of Cambridge University Press.

From Sir Isaac Newton's *Mathematical Principles of Natural Philosophy and His System of the World,* translated into English by Andrew Motte in 1729. The translations revised, and supplied with an historical and explanatory appendix, by Florian Cajori, pp. 393–95. Copyright © 1934 and 1962, The Regents of the University of California.

From Willard V. Quine, "Two Dogmas of Empiricism," in *From a Logical Point of View: Nine Logico-Philosophical Essays* (Cambridge, Mass.: Harvard University Press, 1953), 42–45. Copyright © 1953, 1961, 1980 by the President and Fellows of Harvard College, renewed 1989 by W. V. Quine. Reprinted by permission of the publisher.

From Chen Ning Yang, *Elementary Particles* (Princeton: Princeton University Press, 1962), 53–58. Copyright © 1961 Princeton University Press, 1989 renewed PUP. Reprinted by permission of Princeton University Press.

THEORIES on the
SCRAP HEAP

Introduction

DESCRIPTIVE VERSUS PRESCRIPTIVE PHILOSOPHY OF SCIENCE

One may look at science as an ongoing human activity or as a linguistic entity that records the results of this activity. On either view science involves the description, explanation, and prediction of natural phenomena.

Philosophy of science is an interpretation of science. Its subject matter is evaluative practice within science. The philosopher of science seeks answers to such questions as:

a. What distinguishes a scientific interpretation from a pseudoscientific interpretation?

b. What are the permissible types of scientific explanation?

c. Under what conditions, and to what extent, do observation reports falsify or discredit a theory?

d. When is the replacement of one theory by another justified?

This work investigates the latter two questions.

There are two ways to set about obtaining answers to such questions. The philosopher of science may take either a descriptive approach or a prescriptive approach to the discipline.

Philosophers of science who select the descriptive approach seek to uncover the evaluative standards implicit in scientific practice. These standards may differ from the standards that scientists claim to have applied. The descriptive approach thus may require a certain amount of detective work. A case in point is Darwin's claim that he worked "on true Baconian principles, and without any theory collected facts on a wholesale scale."[1] However, Darwin was not a mere collector of facts with no interest in theories. On the contrary, he was interested in biological and geological hypotheses and he undertook observations with these hypotheses in mind.[2]

Philosophers of science who select the prescriptive approach are not content merely to uncover the evaluative standards that have informed scientific practice. The prescriptivist philosopher of science seeks to formulate evaluative standards by which scientific theories and explanatory arguments ought to be evaluated. Application of these standards is assumed to be important for the creation of good science, and these standards are recommended to practicing scientists.

On the prescriptivist view it is not enough, for instance, to reconstruct Harvey's evaluation of Galen's theory of the blood, Whewell's evaluation of the theory of organic evolution, and Mach's evaluation of atomic theory. The philosopher of science must ask not only "what evaluative procedures have been practiced" but also "what evaluative procedures ought to be utilized to ensure scientific progress."

Since prescriptive standards stipulate how evaluative decisions ought to be made, the failure of specific historical episodes to conform to the requirements of "good practice" need not invalidate the standards. On the other hand, the recommended prescriptive standards must fit *some* important developments. (What counts as a "fit" needs to be specified.) Otherwise, there would be no reason to believe that the standards are appropriate to govern evaluative practice in science.

Thus there is an asymmetry between the descriptive and prescriptive approaches. Descriptive philosophy of science may be pursued without making prescriptive claims. By contrast, the justification of prescriptive standards requires reference to descriptive philosophy of science. To judge whether applications of a prescriptive standard would be consistent with "scientific practice at its best" is to appeal to the results of descriptive analysis.

PLAN OF THE WORK

The present study is primarily a contribution to descriptive philosophy of science, although one or two normative pronouncements may have crept in (particularly in chapter 13). I have included comments on evaluative practice by past and present scientists and philosophers of science. These comments may serve as resource materials on the history of evaluative practice in science. By including excerpts from the writings of scientists and philosophers I hope to achieve some of the benefits that would be realized in a compilation of case studies of scientific evaluative practice. The selection of excerpts no doubt reflects my prejudices about what is important in philosophy of science. However, I hope that I have avoided the pitfall of selecting episodes solely to illustrate prior convictions about scientific method.

I have set the excerpts within an analytical framework developed as answers to a sequence of questions about falsification, rejection, and theory replacement. By so doing I hope to create a narrative continuity that would not be present in a mere collection of excerpts. The procedure I have followed is to survey the responses of scientists and philosophers to the following sequence of questions:

1. Is the popular view that "man proposes, nature disposes" a caricature of science?

2. Are theories ever falsified directly upon appeal to empirical evidence?

3. If not, what can be said about the conditions under which theories are rejected?

4. Is it true that a theory is rejected only if a viable competing theory is available?

5. Given that theories sometimes are in competition with one another, what determines whether theory replacement is justified?

1. THEORIES AND FALSIFICATION

1. The Logic of Falsification

Falsification is an effective rhetorical strategy. Suppose your opponent defends thesis T. If you can show that T implies Q and Q is false, the opponent has a problem.

Socrates was a master of this technique. In Plato's dialogue *Laches*, for instance, the general maintained that "courage" means "persistence in striving for one's goals in the face of opposition." (I have unpacked Laches' identification of "courage" and "endurance.") Socrates replied that if that understanding of courage is correct then foolhardy or reckless acts of persistence qualify as "courageous," and that such acts clearly are not courageous. Laches agreed and withdrew his thesis that identifies courage and persistence.

This falsification strategy implements the *modus tollens* relation of deductive logic:

$$\text{If } P \text{ then } Q$$
$$\underline{\text{Not } Q}$$
$$\text{Therefore, not } P$$

Since *modus tollens* is a valid deduction argument form, it is rational to reject the conclusion "not P" only if one can show that one, or both, of the premises are not true.

Applications of this falsification strategy are widespread within science, even in the early stages of its development. Aristotle, for example, dismissed the suggestion of Herodotus that female fish conceive by swallowing the milt produced by males. He noted that if Herodotus's hypothesis is true then there is a passageway from mouth to uterus. Aristotle maintained that dissections reveal that there is no such passage.[1]

Aristotle also applied the falsification strategy to Empedocles' hypothesis that semen that enters a hot womb produces male offspring, whereas semen that enters a cold womb produces a female offspring. Aristotle pointed out that if this hypothesis is true, then twins conceived in the same womb are both males or both females. He noted, however, that there do exist twins one of which is a male and one of which is a female.[2]

The falsification strategy is particularly effective when a scientist performs an experiment to show that a consequence of a hypothesis is false. An impressive example is Stephen Hales's experiment against the hypothesis of a circulation of sap in plants. According to the circulation hypothesis, sap ascends in the inner core of a plant's stem and descends just inside the outer coating or bark. Hales reasoned that if this hypothesis is correct and a vertical section of the outer coating is removed, then its upper edge should become moist before the lower edge. Hales cut away a three-inch length of bark from an apple branch, placed the branch in water and observed that it was the lower edge of the cut that first became moist.[3]

HALES'S TREE SAP EXPERIMENT

Experiment XLIII

August 20th, at 1 p.m. I took an Apple-branch b (Fig. 26), nine feet long, 1 + ¾ inch diameter, with proportional lateral branches, I cemented it fast to the tube a, by means of the leaden Syphon 1: but first I cut away the bark, and last year's ringlet of wood, for 3 inches length to r. I then filled the tube with water, which was 12 feet long, and ½ inch diameter, having first cut a gap at y through the bark, and last year's wood, 12 inches from the lower end of the stem: the water was very freely imbibed, viz. at the rate of 3 + ½ inches in a minute. In half an hour's time I could plainly perceive the lower part of the gap y to be moister than before; when at the same time the upper part of the wound looked white and dry.

Now in this case the water must necessarily ascend from the tube, thro' the innermost wood because the last year's wood was cut away, for 3 inches length all round the stem; and consequently, if the sap in its natural course descended by the last year's ringlet of wood, and between that and the bark (as many have thought) the water should have descended by the last year's wood, or the bark, and so have first moistened the upper part of the gap y; but on the contrary, the lower part was moisten'd and not the upper part.

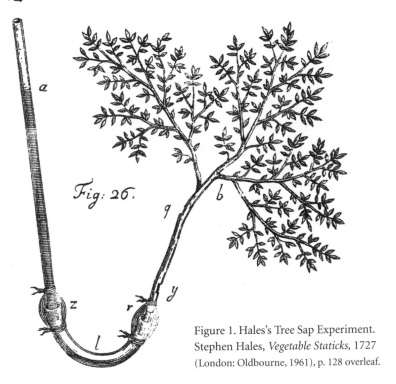

Pl ·12

a

Fig: 26.

b

q

z   *r*   *y*

l

Figure 1. Hales's Tree Sap Experiment.
Stephen Hales, *Vegetable Staticks*, 1727
(London: Oldbourne, 1961), p. 128 overleaf.

I repeated this Experiment with a large Duke-Cherry branch, but could not perceive more moisture at the upper, than the lower part of the gap, which ought to have been, if the sap descends by the last year's wood or the bark.

It was the same in a Quince-branch as the Duke-Cherry.

Hales's application of the falsification strategy is straightforward. The circulation hypothesis implies that the upper edge of the cut becomes moist before the lower edge. Direct unaided observation shows that this does not happen. In more complex cases, falsification is achieved only upon theoretical interpretation of observational evidence.

Count Rumford's cannon-boring experiment is another straightforward attempt at falsification.[4] The experiment was directed against the caloric hypothesis that takes heat to be a substance—an invisible fluid present within bodies. Caloric had been included, along with oxygen, carbon, and iron, as an element in Lavoisier's new system of chemistry (1789). The caloric hypothesis enjoyed widespread support during the late 18th and early 19th centuries.

Rumford observed that if heat is a fluid then there is a limit to the amount of heat that can be produced by a mechanical process that generates friction. Rumford showed that the process of boring a cannon barrel produces a seemingly unlimited amount of heat. When conducted within a box containing water the boring process steadily raises the temperature of the enclosing water and eventually boils it.

RUMFORD'S CANNON-BORING EXPERIMENTS

Being engaged lately in superintending the boring of cannon in the workshops of the military arsenal at Munich, I was struck with the very considerable degree of Heat which a brass gun acquires in a short time in being bored, and with the still more intense Heat (much greater than that of boiling water, as I found by experiment) of the metallic chips separated from it by the borer.

The more I meditated on these phænomena, the more they appeared to me to be curious and interesting. A thorough investigation of them seemed even to bid fair to give a farther insight into the hidden nature of Heat and to enable us to form some reasonable conjectures respecting the existence, or non-existence, of an *igneous fluid*—a subject on which the opinions of philosophers have in all ages been much divided.

In order that the Society may have clear and distinct ideas of the speculations and reasonings to which these appearances gave rise in my mind, and also of the specific objects of philosophical investigation they suggested to me, I must beg leave to state them at some length, and in such manner as I shall think best suited to answer this purpose.

From *whence comes* the Heat actually produced in the mechanical operation above mentioned? Is it furnished by the metallic chips which are separated by the borer from the solid mass of metal?

If this were the case, then, according to the modern doctrines of latent Heat, and of caloric, the *capacity for Heat* of the parts of the metal, so reduced to chips, ought not only to be changed, but the change undergone by them should be sufficiently great to account for all the Heat produced.

But no such change had taken place; for I found, upon taking equal quantities, by weight, of these chips, and of thin slips of the same block of metal separated by means of a fine saw, and putting them at the same temperature (that of boiling water) into equal quantities of cold water (that is to say, at the temperature of 59½° F), the portion of water into which the chips were put was not, to all appearance, heated either less or more than the other portion in which the slips of metal were put.

This experiment being repeated several times, the results were always so nearly the same that I could not determine whether any, or what change had been produced in the metal, in regard to its capacity for Heat, by being reduced to chips by the borer.

From hence it is evident that the Heat produced could not possibly have been furnished at the expence of the latent Heat of the metallic chips.

* * *

The hollow cylinder having been previously cleaned out, and the inside of its bore wiped with a clean towel till it was quite dry, the square iron bar, with the blunt steel borer fixed to the end of it, was put into its place; the mouth of the bore of the cylinder being closed at the same time by means of the circular piston, through the center of which the iron bar passed.

This being done, the box was put in its place, and the joinings of the iron rod and of the neck of the cylinder with the two ends of the box having been made watertight by means of collars of oiled leather, the box was filled with cold water (viz. at the temperature of 60°), and the machine was put in motion.

The result of this beautiful experiment was very striking, and the pleasure it afforded me amply repaid me for all the trouble I had in contriving and arranging the complicated machinery used in making it.

The cylinder, revolving at the rate of about 32 times in a minute, had been in motion but a short time, when I perceived, by putting my hand into the water and touching the outside of the cylinder, that Heat was generated; and it was not long before the water which surrounded the cylinder began to be sensibly warm.

At the end of 1 hour I found, by plunging a thermometer into the water in the box (the quantity of which fluid amounted to 18.77 lb., avoirdupois, or 2¼ wine gallons), that its temperature had been raised no less than 47 degrees; being now 107° of Fahrenheit's scale.

When 30 minutes more had elapsed, or 1 hour and 30 minutes after the machinery had been put in motion, the Heat of the water in the box was 142°.

At the end of 2 hours, reckoning from the beginning of the experiment, the temperature of the water was found to be raised to 178°.

At 2 hours 20 minutes it was at 200°; and at 2 hours 30 minutes it ACTUALLY BOILED!

It would be difficult to describe the surprise and astonishment expressed in the countenances of the bystanders, on seeing so large a quantity of cold water heated, and actually made to boil, without any fire.

* * *

By meditating on the results of all these experiments, we are naturally brought to that great question which has so often been the subject of speculation among philosophers; namely, What is Heat? Is there any such thing as an *igneous fluid?* Is there anything that can with propriety be called *caloric?*

We have seen that a very considerable quantity of Heat may be excited in the friction of two metallic surfaces, and given off in a constant stream or flux *in all directions* without interruption or intermission, and without any signs of diminution or exhaustion.

From whence came the Heat which was continually given off in this manner in the foregoing experiments? Was it furnished by the small particles of metal, detached from the larger solid masses, on their being rubbed together? This, as we have already seen, could not possibly have been the case.

Was it furnished by the air? This could not have been the case; for, in three of the experiments, the machinery being kept immersed in water, the access of the air of the atmosphere was completely prevented.

Was it furnished by the water which surrounded the machinery? That this could not have been the case is evident: first, because this water was continually *receiving Heat* from the machinery, and could not at the same time be *giving to,* and *receiving Heat from,* the same body; and, secondly, because there was no chemical decomposition of any part of this water. Had any such decomposition taken place (which, indeed, could not reasonably have been expected), one of its component elastic fluids (most probably inflammable air) must at the same time have been set at liberty, and, in making its escape into the atmosphere, would have been detected; but though I frequently examined the water to see if any air-bubbles rose up through it, and had even made preparations for catching them, in order to examine them, if any should appear, I could perceive none; nor was there any sign of decomposition of any kind whatever, or other chemical process, going on in the water.

Is it possible that the Heat could have been supplied by means of the iron bar to the end of which the blunt steel borer was fixed? or by the small neck of gun-metal by which the hollow cylinder was united to the cannon? These suppositions appear more improbable even than either of those before mentioned; for Heat was continually going off, or *out of the machinery,* by both these passages, during the whole time the experiment lasted.

And, in reasoning on this subject, we must not forget to consider that most remarkable circumstance, that the source of the Heat generated by friction, in these experiments, appeared evidently to be *inexhaustible.*

It is hardly necessary to add, that anything which any insulated body, or system of bodies, can continue to furnish without limitation, cannot possibly be a material substance; and it appears to me to be extremely difficult, if not quite impossible, to form any distinct idea of anything capable of being excited and communicated in the manner the Heat was excited and communicated in these experiments, except it be MOTION.

Rumford's experiments were not a decisive refutation of the caloric hypothesis. He did not establish that an *inexhaustible* amount of heat is generated. Nevertheless, the very great amount of heat released made implausible the hypothesis that the heat produced is a fluid escaping from the metal.

Decades later, a successful competing theory was formulated. According to the kinetic molecular theory, temperature is a measure of the intensity of molecular motions, and heat is the product of an increase in such motions. Supporters of the kinetic molecular theory sometimes cited Rumford's experiments to show that the "heat is an invisible fluid" theory is not a viable alternative.

In the mid-eighteenth century, Pierre de Maupertuis sought to falsify monoparental theories of heredity.[5] Maupertuis traced the genealogy of a Berlin physician, Jacob Ruhe. Some members of the Ruhe family possessed an extra finger or two (polydactyly). A genealogical chart revealed that this trait is passed to members of the next generation from both male and female parents.

This finding would appear to falsify the then popular theories of ovism and animalculism. Ovism and animalculism are versions of preformationism. Preformationist theories interpret embryological development to be an unfolding of structures already present in the egg or sperm. Ovists (e.g., Swammerdam) held that offspring are encapsulated in the female egg; and animalculists (e.g., Hartsoeker, Dalenpatius) held that off-spring are encapsulated in the male sperm.[6]

MAUPERTUIS'S CONCLUSIONS ABOUT POLYDACTYLY

Jacob Ruhe, surgeon of Berlin, is one of these types. Born with six digits on each hand and each foot, he inherited this peculiarity from his mother Elisabeth Ruhen, who inherited it from her mother Elisabeth Horstmann, of Rostock. Eliza-beth Ruhen transmitted it to four children of eight she had by Jean Christian Ruhe, who had nothing extraordinary about his feet or hands. Jacob Ruhe, one of these six-digited children, espoused, at Dantzig in 1733, Sophie Louise de Thungen who had no extraordinary trait: he had by her six children; two boys were six-digited. One of them, Jacob Ernest, had six digits on the left foot and five on the right: he had on the right hand a sixth finger, which was amputated; on the left he had in the place of the sixth digit only a stump.

One sees from this genealogy which I have followed with exactitude, that polydactyly (six-digitism) is transmitted equally by the father and by the mother: one sees that it is altered through the mating with five-digited persons. Through these repeated matings it must probably disappear (*s'eteindre*); and must be per-petuated through matings in which it is carried in common by both sexes.

Scientists nevertheless may accept Maupertuis's finding without con-ceding that it falsifies preformationism. Perhaps animalculism is correct. Every offspring is an unfolding of rudimentary structures already present in the sperm. Perhaps aberrations such as polydactyly are not hereditary

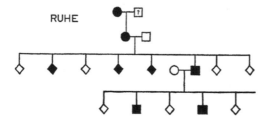

Where O - *normal female*
　　　□ - normal male
　　　◆ - *normal offspring, sex unrecorded*
　　　● - *polydactylous female*

Figure 2. Ruhe Family Genealogy. Elizabeth B. Gasking, *Investigations into Genera-tion, 1651–1828* (London: Hutchinson, 1967), p. 81.

traits at all. Perhaps they are the results of mistakes that occur during embryological development. Perhaps it is just a matter of chance that such mistakes are prevalent in certain families.

Maupertuis himself assessed the probability of such a convergence of chance aberrations and concluded it to be negligible. He declared that

> if one wished to regard the continuation of polydactyly as an effect of pure chance, it would be necessary to see what the probability is that this accidental variation in a first parent would be repeated in his descendants. After a search which I have made in a city which has one hundred thousand inhabitants, I have found two men who had this singularity. Let us suppose, which is difficult, that three others have escaped me; and that in 20,000 men one can reckon on one six-digited: the probability that his son or daughter will not be born with polydactyly at all is 20,000 to 1; and that his son and his grandson will not be six-digited at all is 20,000 × 20,000 or 400,000,000 to 1: finally the probability that this singularity will not continue during three generations would be 8,000,000,000,000 to 1; a number so great that the certainty of the best demonstrated things of physics does not approach these probabilities.[7]

Maupertuis thus dismissed the appeal to chance. Nevertheless, the documentation of the Ruhe family history, together with the elimination of the chance alternative, still does not falsify preformationism. The preformationist can argue that in cases of polydactyly some yet-undiscovered factor (diet, water, temperature . . .) interferes with the normal course of embryological development. On this defense, offspring are an unfolding of structures present in the sperm (or egg), but occasionally (unspecified) external factors alter this unfolding. All that is falsified by the Ruhe family genealogy is that all polydactylous offspring have a polydactylous male parent (or that all such offspring have a polydactylous female parent).

The falsificationist strategy played a large role in the developments that produced a revolution in chemistry.[8] In 1774, French chemist Pierre Bayen announced that the product of heating the red calx of mercury is "fixed air" (CO_2). Antoine Lavoisier rejected Bayen's conclusion. He reasoned that if the decomposition product is fixed air, then this product dissolves in water upon shaking and forms a white precipitate in lime water ($Ca(OH)_2$). Lavoisier demonstrated that neither effect is observed to take place.

1. Priestley's Nitrous Air Test

1 nitrous air + 1 ordinary air = 1.4 residual
(NO)	.2 O_2	.6 NO
	.8 N_2	.8 N_2

volume decrease = $\dfrac{2.0 - 1.4}{2.0}$ = 30%

1 nitrous air + 2 ordinary air = 1.8 residual
(NO)	.4 O_2	.2 O_2
	1.6 N_2	1.6 N_2

volume decrease = $\dfrac{3.0 - 1.8}{3.0}$ = 40%

1 nitrous air + 3 ordinary air = 2.5 residual
(NO)	.6 O_2	.1 O_2
	2.4 N_2	2.4 N_2

volume decrease = $\dfrac{4.0 - 2.5}{4.0}$ = 37.5%

Lavoisier was interested at that time in a test Joseph Priestley had developed for the "goodness" of air. Air quality was a preoccupation of eighteenth-century investigators. It was widely recognized that the quality of air in mines and the holds of ships could deteriorate to life-threatening levels. Priestley's test showed promise as an indicator of air quality. It was based on a reaction between nitrous air and ordinary air, a reaction that produces red fumes that are soluble in water:

$$\text{nitrous air} + \text{ordinary air} = \text{red fumes}$$
$$\text{(NO)} \qquad \text{(N}_2 + \text{O}_2) \qquad \text{(NO}_2)$$

If nitrous air is added to ordinary air over water, the red fumes produced dissolve, leaving a residual gas volume that is less than the initial volume of nitrous air plus ordinary air. Priestley established that a maximum contraction of volume occurs when the initial volume ratio is 1 nitrous air to 2 ordinary air (see table 1 above).

If the air to which a one-half measure of nitrous air is added is not pure, then the volume decrease will be less than 40%. Suppose that before testing, a mouse is allowed to breathe the air, or a candle is burned in that air. The addition of one volume of nitrous air to two volumes of this "diminished air" reduces the initial volume by an amount less than 40%. In the limiting case of "putrid air" in which a mouse has expired or a candle has burned out, there is no decrease in volume:

$$1 \text{ nitrous air (NO)} + 2 \text{ putrid air (N}_2) = 3 \text{ residual (1 NO} + 2 \text{ N}_2)$$

Lavoisier was impressed by the diagnostic power of Priestley's nitrous air test. It occurred to him that he could falsify Bayen's fixed air hypothesis by adding nitrous air to the gas produced by heating the red calx of mercury. He performed the test. The observed volume decrease was slightly greater than the 40% obtained with common air.

The theoretical volume decrease is 50%:

1 nitrous air (NO) + 2 gaseous product (O_2) = 1.5 residual (O_2)
of the decomposition
of the red calx of
mercury

$$\text{volume decrease} = \frac{3.0 - 1.5}{3.0} = 50\%.$$

Had the observed decrease been less than 40%, Lavoisier surely would have investigated further. But it was easy for Lavoisier to attribute the greater than 40% volume to experimental error, since he had no reason to believe that there exists a "super-good air." Lavoisier reported to the French Academy of Science in 1775 that the gaseous product of the decomposition of the red calx of mercury is common air.

Lavoisier's falsification of Bayen's fixed air hypothesis was successful despite the fact that he was wrong about the nature of the residual gas. His common air hypothesis quickly met the same fate as Bayen's hypothesis. Priestley read Lavoisier's report and promptly applied the falsification strategy to the common air hypothesis.

Priestley argued that if the residual gas from the nitrous air test is common air then the addition of more nitrous air does not reduce the percentage volume decrease. As indicated in Table 1, the maximum volume decrease—40%—is achieved at a ratio of one volume of nitrous air to two volumes of common air. At a one-to-one ratio the volume decrease is only 30%. Priestley showed that when nitrous air is added to the gas produced from the decomposition of the red calx of mercury there is a progressive volume decrease up to a ratio of two volumes of nitrous air to one volume of the gas. At the two-to-one ratio, no gas remains after solution of the red fumes.

2 nitrous air (NO) + 1 gas from the = 2 soluble red fumes (NO_2)
decomposition
of the red calx
of mercury
(O_2)

Priestley concluded that the gas produced from the decomposition of the red calx of mercury is not common air. He provided additional support for this conclusion by showing that the mystery gas is superior to common air in supporting combustion and the respiration of a mouse.

PRIESTLEY ON THE "GOODNESS" OF AIR

On the 8th of this month I procured a mouse, and put it into a glass vessel, containing two ounce-measures of the air from *mercurius calcinatus.* Had it been common air, a full-grown mouse, as this was, would have lived about a quarter of an hour. In this air however, my mouse lived a full half hour; and though it was taken out seemingly dead, it appeared to have been only exceedingly chilled; for, upon being held to the fire, it presently revived, and appeared not to have received any harm from the experiment.

By this I was confirmed in my conclusion, that the air extracted from *mercurius calcinatus,* &c. was, *at least, as good* as common air; but I did not certainly conclude that it was any *better;* because, though one mouse would live only a quarter of an hour in a given quantity of air, I knew it was not impossible but that another mouse might have lived in it half an hour; so little accuracy is there in this method of ascertaining the goodness of air: and indeed I have never had recourse to it for my own satisfaction, since the discovery of that most ready, accurate, and elegant test that nitrous air furnishes. But in this case I had a view to publishing the most generally satisfactory account of my experiments that the nature of the thing would admit of.

This experiment with the mouse, when I had reflected upon it some time, gave me so much suspicion that the air into which I had put it was better than common air, that I was induced, the day after, to apply the test of nitrous air to a small part of that very quantity of air which the mouse had breathed so long; so that, had it been common air, I was satisfied it must have been very nearly, if not altogether, as noxious as possible, so as not to be affected by nitrous air; when, to my surprize again, I found that though it had been breathed so long, it was still better than common air. For after mixing it with nitrous air, in the usual proportion of two to one, it was diminished in proportion of $4\frac{1}{2}$ to $3\frac{1}{2}$; that is, the nitrous air had made it two ninths less than before, and this in a very short space of time; whereas I had never found that in the longest time, any common air was reduced more than one fifth of its bulk by any proportion of nitrous air, nor more than one fourth by any phlogistic process whatever. Thinking of this extraordinary fact upon my pillow, the next morning I put another measure of nitrous air to the same mixture, and, to my utter astonishment, found that it was farther diminished to almost one half of its original quantity. I then put a third measure to it; but this did not diminish it any farther: but, however, left it one measure less than it was even after the mouse had been taken out of it.

Being now fully satisfied that this air, even after the mouse had breathed it half an hour, was much better than common air; and having a quantity of it still left, sufficient for the experiment, viz. an ounce-measure and a half, I put the mouse into it; when I observed that it seemed to feel no shock upon being put into it, evident signs of which would have been visible, if the air had not been very wholesome; but that it remained perfectly at its ease another full half hour, when I took it out quite lively and vigorous. Measuring the air the next day, I found it to be reduced from 1½ to ⅔ of an ounce-measure. And after this, if I remember well (for in my register of the day I only find it noted, that it was considerably diminished by nitrous air) it was nearly as good as common air. It was evident, indeed, from the mouse having been taken out quite vigorous, that the air could not have been rendered very noxious.

For my farther satisfaction I procured another mouse, and putting it into less than two ounce-measures of air extracted from *mercurius calcinatus* and air from red precipitate (which, having found them to be of the same quality, I had mixed together) it lived three quarters of an hour. But not having had the precaution to set the vessel in a warm place, I suspect that the mouse died of cold. However, as it had lived three times as long as it could probably have lived in the same quantity of common air, and I did not expect much accuracy from this kind of test, I did not think it necessary to make any more experiments with mice.

Being now fully satisfied of the superior goodness of this kind of air, I proceeded to measure that degree of purity with as much accuracy as I could, by the test of nitrous air; and I began with putting one measure of nitrous air to two measures of this air, as if I had been examining common air; and now I observed that the diminution was evidently greater than common air would have suffered by the same treatment. A second measure of nitrous air reduced it to two thirds of its original quantity, and a third measure to one half. Suspecting that the diminution could not proceed much farther, I then added only half a measure of nitrous air, by which it was diminished still more; but not much, and another half measure made it more than half of its original quantity; so that, in this case, two measures of this air took more than two measures of nitrous air, and yet remained less than half of what it was. Five measures brought it pretty exactly to its original dimensions.

At the same time, air from the red *precipitate* was diminished in the same proportion as that from *mercurius calcinatus,* five measures of nitrous air being received by two measures of this without any increase of dimensions. Now as common air takes about one half of its bulk of nitrous air, before it begins to receive any addition to its dimensions from more nitrous air, and this air took more than four half-measures before it ceased to be diminished by more nitrous air, and even five half-measures made no addition to its original dimensions, I conclude that it was between four and five times as good common air. It will be seen that I have since procured air better than this, even between five and six times as good as the best common air that I have ever met with.

Being now fully satisfied with respect to the nature of this new species of air, viz. that, being capable of taking more phlogiston from nitrous air, it therefore originally contains less of this principle; my next inquiry was, by what means it comes to be so pure, or philosophically speaking, to be so much *dephlogisticated.*[9]

Priestley's falsification of Lavoisier's common air hypothesis is successful even though his theory about the mystery gas is incorrect. Priestley was committed to the phlogiston theory. According to this theory, the burning of metals releases an invisible gaseous substance called phlogiston:

$$\text{zinc (calx of zinc} + \varphi) = \text{calx of zinc} + \varphi \uparrow$$
$$\Delta$$

On the phlogiston theory, the reduction of a calx requires the presence of phlogiston:

$$\text{calx of zinc} + \text{charcoal (rich in } \varphi) = \text{zinc (calx} + \varphi)$$
$$\Delta$$

However, the red calx of mercury yields the metal directly upon heating without the presence of charcoal or wood. On the phlogiston theory, the metal mercury is a combination of its calx plus phlogiston.

It seemed to Priestley that the required phlogiston could have come only from the air. Hence he interpreted the gaseous product of the mercury-calx decomposition to be air from which phlogiston had been removed, or "dephlogisticated" air:

red calx of + air (contains φ) = mercury + dephlogisticated air
mercury   (calx + φ)   (air − φ)

Lavoisier abandoned his common air hypothesis upon reading Priestley's report on the properties of the gaseous product of the red-calx decomposition. He agreed with Priestley that this gas supported combustion and respiration better than did common air. However, he reversed Priestley's phlogiston theory interpretation by taking metals to be elementary substances and calxes to be compounds of a metal and oxygen. Lavoisier maintained that when a metal is heated to form a calx, oxygen is withdrawn from the air and combines with the metal. Conversely, when a calx is heated, usually in the presence of charcoal, oxygen is released. (The red calx of mercury is an exception. It releases oxygen upon moderate heating without the presence of charcoal.)

An impressive application of an experimental result to falsify a previously formulated theory took place in Ernest Rutherford's Manchester laboratory in 1909. J. J. Thomson had suggested that atoms are spheres of positive charge in which negatively charged electrons are embedded. Thomson's "plum pudding" model could be applied to picture atoms of the elements that make up the periodic table. For instance, see figure 3.

Hans Geiger observed that most α-particles in a narrowly focused beam pass through thin foils of gold or silver with negligible deflection. He determined that the most probable angle of deflection for a particle in the beam is about one degree. This was not a surprising result, given the high speed and mass of the particles (α-particles are nuclei of helium atoms).

Rutherford then suggested that Geiger's associate Ernest Marsden in-

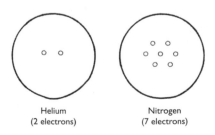

Helium
(2 electrons)

Nitrogen
(7 electrons)

Figure 3. Thomson's "Plum Pudding" Model of Atoms.

vestigate the scattering process to see if any particles are deflected through large angles. Rutherford recalled that he was surprised that Marsden's data revealed an occasional large deflection. Indeed, one particle in 20,000 is scattered by thin gold foil through an angle greater than ninety degrees.

Rutherford recognized that this result falsified Thomson's model, because that model implies that there should be only small deflections caused by electrostatic repulsion between α-particles and the positively charged "pudding" of the atoms. (An embedded electron is roughly 7,000 times lighter than an α-particle and would not appreciably deflect the particle by electrostatic attraction.) In response to the experimental findings, Rutherford formulated a "nuclear model" of the atom, in which a tiny dense positively charged nucleus is surrounded by electrons.[10]

RUTHERFORD ON ALPHA-PARTICLE SCATTERING

In the early days I had observed the scattering of α-particles, and Dr. Geiger in my laboratory had examined it in detail. He found, in thin pieces of heavy metal, that the scattering was usually small, of the order of one degree. One day Geiger came to me and said, "Don't you think that young Marsden, whom I am training in radioactive methods, ought to begin a small research?" Now I had thought that too, so I said, "Why not let him see if any α-particles can be scattered through a large angle?" I may tell you in confidence that I did not believe that they would be, since we knew that the α-particle was a very fast massive particle, with a great deal of energy, and you could show that if the scattering was due to the accumulated effect of a number of small scatterings the chance of an α-particle's being scattered backwards was very small. Then I remember two or three days later Geiger coming to me in great excitement and saying, "We have been able to get some of the α-particles coming backwards. . . ." It was quite the most incredible event that has ever happened to me in my life. It was almost as incredible as if you fired a 15-inch shell at a piece of tissue paper and it came back and hit you. On consideration I realized that this scattering backwards must be the result of a single collision, and when I made calculations I saw that it was impossible to get anything of that order of magnitude unless you took a system in which the greater part of the mass of the atom was concentrated in a minute nucleus. It was then that I had the idea of an atom with a minute massive centre carrying a charge. I worked out mathematically what laws the scattering should obey, and I found that the number of particles scattered through a given angle should be proportional to the thickness of the scattering foil, the square of the nuclear charge, and inversely proportional to the fourth power of the velocity. These deductions were later verified by Geiger and Marsden in a series of beautiful experiments.

Now let us consider what deductions could be made at that stage. By considering how close to the nucleus the α-particles could go, and yet be scattered normally, I could show that the size of the nucleus must be very small. I also estimated the magnitude of the charge and made it about a hundred times as great as the electronic charge e. It was not possible to make an accurate estimate, but general evidence indicated that the nucleus of hydrogen must have a charge e, helium 2e, and so on. Geiger and Marsden examined the scattering in different elements and found that the amount of scattering varied as the square of the atomic weight. This result was rough but quite sufficient: it indicated that the charge on a nucleus was roughly proportional to the atomic weight.

Hales, Rumford, Priestley, and Marsden achieved falsification by first noting that hypothesis H materially implies consequent Q and then performing an experiment to show that Q is not the case. This is the usual sequence. It conforms to the "man proposes, nature disposes" image of scientific inquiry.

However, there is nothing about the logic of falsification that dictates that $(H \supset Q)$ be formulated first and ~Q be determined subsequently. In some cases in which scientists achieve falsification, ~Q is known first and the material implication $(H \supset Q)$ is established only subsequently. An important example is Isaac Newton's argument against Descartes' vortex theory of the solar system.

Newton and his rivals antecedently accepted the truth of Kepler's laws. The third law states that the period (T) of a planet is proportional to the three halves power of its mean distance from the sun. For any two planets the following proportionality holds:

$$(T_1/T_2) = (D_1/D_2)^{3/2}$$

Newton argued that if the planets are carried in stable orbits around the sun by an invisible ethereal whirlpool, then the density of the vortex at each planet's distance from the center is equal to that of the planet itself. If the vortex fluid at a planet's location were more dense than the planet then the planet would spiral in toward the center of the vortex. And if the vortex fluid at a planet's location were less dense than the planet then the planet would recede from the center. Newton thought it implausible that an invisible vortex of such high (and varying) density exists. But he was willing to concede to the Cartesians that there might be such a vortex because he could show that the existence of such a vortex is inconsistent with Kepler's third law. Newton established that a planet could remain in a stable orbit in a Cartesian vortex only if its period is proportional to the square of its distance from the center. Since this result contradicts Kepler's third law, Descartes' vortex hypothesis is false.[11]

NEWTON ON THE VORTEX THEORY

I have endeavored in this Proposition, to investigate the properties of vortices, that I might find whether the celestial phenomena can be explained by them; for the phenomenon is this, that the periodic times of the planets revolving about Jupiter are as the 3/2th power of their distances from Jupiter's centre; and the same rule obtains also among the planets that revolve about the sun. And these rules obtain also with the greatest accuracy, as far as has been yet discovered by astronomical observation. Therefore if those planets are carried round in vortices revolving about Jupiter and the sun, the vortices must revolve according to that law. But here we found the periodic times of the parts of the vortex to be as the square of the distances from the centre of motion; and this ratio cannot be diminished and reduced to the 3/2th power, unless either the matter of the vortex be more fluid the farther it is from the centre, or the resistance arising from the want of lubricity in the parts of the fluid should, as the velocity with which the parts of the fluid are separated goes on increasing, be augmented with it in a greater ratio than that in which the velocity increases. But neither of these suppositions seem reasonable. The more gross and less fluid parts will tend to the circumference, unless they are heavy towards the centre. And though, for the sake of demonstration, I proposed, at the beginning of this Section, an Hypothesis that the resistance is proportional to the velocity, nevertheless, it is in truth probably that the resistance is in a less ratio than that of the velocity; which granted, the periodic times of the parts of the vortex will be in a greater ratio than the square of the distances from its centre. If, as some think, the vortices move more swiftly near the centre, then slower to a certain limit, then again swifter near the circumference, certainly neither the 3/2th power, nor any other certain and determinate power, can obtain in them. Let philosophers then see how that phenomenon of the 3/2th power can be accounted for by vortices.

Proposition LIII, Theorem XLI

Bodies carried about in a vortex, and returning in the same orbit, are of the same density with the vortex, and are moved according to the same law with the parts of the vortex, as to velocity and direction of motion.

For if any small part of the vortex, whose particles or physical points continue a given situation among themselves, be supposed to be congealed, this particle will move according to the same law as before, since no change is made either in its density, inertia, or figure. And again; if a congealed or solid part of the vortex be of the same density with the rest of the vortex, and be resolved into a fluid, this will move according to the same law as before, except so far as its particles, now become fluid, may be moved among themselves. Neglect, therefore, the motion of the particles among themselves as not at all concerning the progressive motion of the whole, and the motion of the whole will be the same as before. But this motion will be the same with the motion of other parts of the vortex at equal distances from the centre; because the solid, now resolved into a fluid, is become exactly like the other parts of the vortex. Therefore, a solid, if it be of the same density with the matter of the vortex, will move with the same motion as the parts thereof, being relatively at rest in the matter that surrounds it. If it be more dense, it will endeavor more than before to recede from the centre; and therefore overcoming that force of the vortex, by which, being, as it were, kept in equilibrium, it was retained in its orbit, it will recede from the centre, and in its revolution describe a spiral, returning no longer into the same orbit. And, by the same argument, if it be more rare, it will approach to the centre. Therefore it can never continually go round in the same orbit, unless it be of the same density with the fluid. But we have shown in that case that it would revolve according to the same law with those parts of the fluid that are at the same or equal distances from the centre of the vortex.

Vortex theorists such as Leibniz and De Molieres conceded that Newton's argument was decisive against the single sun-centered vortex postulated by Descartes. They maintained, however, that theories about the interaction of multiple vortices can be developed that are consistent with Kepler's laws.[12]

2. The Limits of Falsification

The circulation of sap hypothesis, the common air hypothesis, and the plum pudding model are falsified only if the statements about the relevant experiments are true. This is a matter of logic. The *modus tollens* argument form, $[(H \supset Q)$ & $\sim Q]/ \therefore \sim H$, states only that "it is necessary that if '$H \supset Q$' is true and '$\sim Q$' is true, then '$\sim H$' is true as well." If '$\sim Q$' is not true, then no reason has been provided to take H to be false. This point of logic was emphasized by Pierre Duhem.

Pierre Duhem set forth the limitations of the falsification procedure in *The Aim and Structure of Physical Theory* (1906). He emphasized that what is falsified by an observation report is only a conjunction of premises.[1] A simple case is an experimental result inconsistent with a prediction derived from a generalization. Consider the law "all blue litmus turns red in acid solution." We predict that the paper turns red on the basis of the following argument:

P_1—All pieces of blue litmus paper, placed in an acid solution, turn red
P_2—This piece of blue litmus paper is placed in an acid solution
C—Therefore this piece of blue litmus paper turns red.

If the litmus paper does not turn red, then one or more of the premises is false. The generalization itself is not falsified by the negative result. It may be false instead that there is blue litmus dye on the paper. And it may be false as well that the solution in question is acidic. The logic of the situation is represented by the following:

$$(P_1 \& P_2) \supset C$$
$$\underline{\sim C}$$
$$\therefore \sim (P_1 \& P_2)$$

Duhem pointed out that the usual context of falsification is still more complex. To derive a prediction about the occurrence of a phenomenon the scientist typically presupposes a number of individual hypotheses, among them hypotheses about the operation of measuring instruments. Given a *prima facie* falsifying instance, the theorist may accommodate the instance by abandoning or modifying any of the hypotheses used to make the prediction. She also may retain all hypotheses involved and reject instead the statement about the relevant conditions under which the test was performed. Duhem insisted that no observation report ever is decisive against a scientific hypothesis.[2]

DUHEM ON THE LOGIC OF FALSIFICATION

An experiment in physics can never condemn an isolated hypothesis but only a whole theoretical group

The physicist who carries out an experiment, or gives a report of one, implicitly recognizes the accuracy of a whole group of theories. Let us accept this principle and see what consequences we may deduce from it when we seek to estimate the role and logical import of a physical experiment.

In order to avoid any confusion we shall distinguish two sorts of experiments: experiments of *application,* which we shall first just mention, and experiments of *testing,* which will be our chief concern.

*   *   *

A physicist disputes a certain law; he calls into doubt a certain theoretical point. How will he justify these doubts? How will he demonstrate the inaccuracy of the law? From the proposition under indictment he will derive the prediction of an experimental fact; he will bring into existence the conditions under which this fact should be produced; if the predicted fact is not produced, the proposition which served as the basis of the prediction will be irremediably condemned.

*   *   *

A physicist decides to demonstrate the inaccuracy of a proposition; in order to deduce from this proposition the prediction of a phenomenon and institute the experiment which is to show whether this phenomenon is or is not produced, in order to interpret the results of this experiment and establish that the predicted phenomenon is not produced, he does not confine himself to making use of the proposition in question; he makes use also of a whole group of theories accepted by him as beyond dispute. The prediction of the phenomenon, whose nonproduction is to cut off debate, does not derive from the proposition challenged if taken by itself, but from the proposition at issue joined to that whole group of theories; if the predicted phenomenon is not produced, not only is the proposition questioned at fault, but so is the whole theoretical scaffolding used by the physicist. The only thing the experiment teaches us is that among the propositions used to predict the phenomenon and to establish whether it would be produced, there is at least one error; but where this error lies is just what it does not tell us. The physicist may declare that this error is contained in exactly the proposition he wishes to refute, but is he sure it is not in another proposition? If he is, he accepts implicitly the accuracy of all the other propositions he has used, and the validity of his conclusion is as great as the validity of his confidence.

* * *

We know that Newton conceived the emission theory for optical phenomena. The emission theory supposes light to be formed of extremely thin projectiles, thrown out with very great speed by the sun and other sources of light; these projectiles penetrate all transparent bodies; on account of the various parts of the media through which they move, they undergo attractions and repulsions; when the distance separating the acting particles is very small these actions are very powerful, and they vanish when the masses between which they act are appreciably far from each other. These essential hypotheses joined to several others, which we pass over without mention, lead to the formulation of a complete theory of reflection and refraction of light; in particular, they imply the following proposition: The index of refraction of light passing from one medium into another is equal to the velocity of the light projectile within the medium it penetrates, divided by the velocity of the same projectile in the medium it leaves behind.

This is the proposition that Arago chose in order to show that the theory of emission is in contradiction with the facts. From this proposition a second follows: Light travels faster in water than in air. Now Arago had indicated an appropriate procedure for comparing the velocity of light in air with the velocity of light in water; the procedure, it is true, was inapplicable, but Foucault modified the experiment in such a way that it could be carried out; he found that the light was propagated less rapidly in water than in air. We may conclude from this, with Foucault, that the system of emission is incompatible with the facts.

I say the system of emission and not the hypothesis of emission; in fact, what the experiment declares stained with error is the whole group of propositions accepted by Newton, and after him by Laplace and Biot, that is, the whole theory from which we deduce the relation between the index of refraction and the velocity of light in various media. But in condemning this system as a whole by declaring it stained with error, the experiment does not tell us where the error lies. Is it in the fundamental hypothesis that light consists in projectiles thrown out with great speed by luminous bodies? Is it in some other assumption concerning the actions experienced by light corpuscles due to the media through which they move? We know nothing about that. It would be rash to believe, as Arago seems to have thought, that Foucault's experiment condemns once and for all the very hypothesis of emission, i.e., the assimilation of a ray of light to a swarm of projectiles. If physicists had attached some value to this task, they would undoubtedly have succeeded in founding on this assumption a system of optics that would agree with Foucault's experiment.

In sum, the physicist can never subject an isolated hypothesis to experimental test, but only a whole group of hypotheses; when the experiment is in disagreement with his predictions, what he learns is that at least one of the hypotheses constituting this group is unacceptable and ought to be modified; but the experiment does not designate which one should be changed.

Duhem's pessimistic conclusion about falsification is perhaps too strong. Duhem is correct that observation report $\sim\Psi a$ alone does not falsify generalization $(\forall x) (\Phi x \supset \Psi x)$. But a generalization of the form $(\forall x) \Phi x$ is falsified by observation report $\sim\Phi a$. That "everything is spherical" is falsified by a description of the Washington monument.

Unfortunately, scientists are not much interested in generalizations of the form $(\forall x) \Phi x$. They are much more interested in singly (or multiply) quantified *conditional* generalizations of the form $(\forall x) (\Phi x \supset \Psi x)$. Examples include the following:

"For all x, if x is a sodium sample, then x reacts with chlorine";
"For all x, if x is the child of blue-eyed parents, then x has blue eyes"; and
"For all x, if x is a body moving in the absence of impressed forces, then x continues in constant straight-line motion."

To falsify a universal conditional, $(\forall x) (\Phi x \supset \Psi x)$, it is necessary to establish for a particular instance a, both Φa and $\sim\Psi a$. Since one may be wrong about the supposed truth of Φa, it does not follow that because $\sim\Psi a$ is observed, the generalization must be false. As Duhem emphasized, a scientific law is not falsified by a single observation report.

Willard van Orman Quine noted that since no experimental result is ever decisive against a single hypothesis, a negative test result can be accommodated by making changes elsewhere in the system of hypotheses and statements about relevant conditions. Suppose theory T, in conjunction with auxiliary hypotheses H and statement of relevant conditions C, implies observation report O:

$$T \,\&\, A$$
$$\underline{\quad C \quad}$$
$$\therefore O$$

If what is observed is $\sim O$, theory T still may be retained. Quine emphasized that one always may substitute a different auxiliary hypothesis A^*, such that the following holds true, thereby shielding T from falsification.

$$T \,\&\, A^*$$
$$\underline{\quad C \quad}$$
$$\therefore \sim O$$

He granted that A^* may be complex, or otherwise implausible, but one always can find an A^* that protects T from falsification.

QUINE'S "FIELD OF FORCE" IMAGE

Total science is like a field of force whose boundary conditions are experience. A conflict with experience at the periphery occasions readjustments in the interior of the field, Truth values have to be redistributed over some of our statements. Reevaluation of some statements entails reevaluation of others, because of their logical interconnections—the logical laws being in turn simply certain further statements of the system, certain further elements of the field. Having reevaluated one statement we must reevaluate some others, which may be statements logically connected with the first or may be the statements of logical connections themselves. But the total field is so underdetermined by its boundary conditions, experience, that there is much latitude of choice as to what statements to reevaluate in the light of any single contrary experience. No particular experiences are linked with any particular statements in the interior of the field, except indirectly through considerations of equilibrium affecting the field as a whole.

If this view is right, it is misleading to speak of the empirical content of an individual statement especially if it is a statement at all remote from the experiential periphery of the field. Furthermore it becomes folly to seek a boundary between synthetic statements, which hold contingently on experience, and analytic statements, which hold come what may. Any statement can be held true come what may, if we make drastic enough adjustments elsewhere in the system. Even a statement very close to the periphery can be held true in the face of recalcitrant experience by pleading hallucination or by amending certain statements of the kind called logical laws. Conversely, by the same token, no statement is immune to revision. Revision even of the logical law of the excluded middle has been proposed as a means of simplifying quantum mechanics; and what difference is there in principle between such a shift and the shift whereby Kepler superseded Ptolemy, or Einstein Newton, or Darwin Aristotle?

A recalcitrant experience can, I have urged, be accommodated by any of various alternative reevaluations in various alternative quarters of the total system; but in the cases which we are now imagining, our natural tendency to disturb the total system as little as possible would lead us to focus our revisions upon these specific statements concerning brick houses or centaurs. These statements are felt, therefore, to have a sharper empirical reference than highly theoretical statements of physics or logic or ontology. The latter statements may be thought of as relatively centrally located within the total network, meaning merely that little preferential connection with any particular sense data obtrudes itself.

As an empiricist I continue to think of the conceptual scheme of science as a tool, ultimately, for predicting future experience in the light of past experience. Physical objects are conceptually imported into the situation as convenient intermediaries—not by definition in terms of experience, but simply as irreducible posits comparable, epistemologically, to the gods of Homer. For my part I do, qua lay physicist, believe in physical objects and not in Homer's gods; and I consider it a scientific error to believe otherwise. But in point of epistemological footing the physical objects and the gods differ only in degree and not in kind. Both sorts of entities enter our conception only as cultural posits. The myth of physical objects is epistemologically superior to most in that it has proved more efficacious than other myths as a device for working a manageable structure into the flux of experience.

Positing does not stop with macroscopic physical objects. Objects at the atomic level are posited to make the laws of macroscopic objects, and ultimately the laws of experience, simpler and more manageable; and we need not expect or demand full definition of atomic and subatomic entities in terms of macroscopic ones, any more than definition of macroscopic things in terms of sense data. Science is a continuation of common sense, and it continues the common-sense expedient of swelling ontology to simplify theory.

Physical objects, small and large, are not the only posits. Forces are another example; and indeed we are told nowadays that the boundary between energy and matter is obsolete. Moreover, the abstract entities which are the substance of mathematics—ultimately classes and classes of classes and so on up—are another posit in the same spirit. Epistemologically these are on the same footing with physical objects and gods, neither better nor worse except for differences, in the degree to which they expedite our dealings with sense experiences.[3]

A historical episode that fits Quine's "field of force" image rather well is the case of *Megaloceros,* the "Irish elk." *Megaloceros* posed a problem for the theory of evolution by natural selection. Dobzhansky put the problem as follows: "Starting with late tertiary time (Pliocene) the ancestors of *Megaloceros* were getting progressively larger in size, and, as they were getting larger, the antlers in the males were getting more and more enormous, until during the Ice Age the antlers reached seemingly absurd dimensions; and finally these animals died out. Some paleontologists surmised that such huge antlers must have been injurious to their carriers, and concluded that *Megaloceros* died out *because* its antlers got too big for it to carry. In other words, the evolutionary trend towards large antlers developed such a momentum that it could not stop, even when it got to be harmful and led to extinction."[4]

The increase in antler size would seem to be nonadaptive. The species became extinct, probably shortly after reforestation followed upon the retreat of glaciers.[5] The top-heavy elk, carrying ninety-pound antlers on a five-pound skull, did not achieve success as a forest dweller.

The persistence of a nonadaptive trait, such as the increasing antler size of *Megaloceros,* is *prima facie* disconfirming evidence for the theory of natural selection. The theory of natural selection attributes evolutionary change to selective pressure that confers differential reproductive advan-

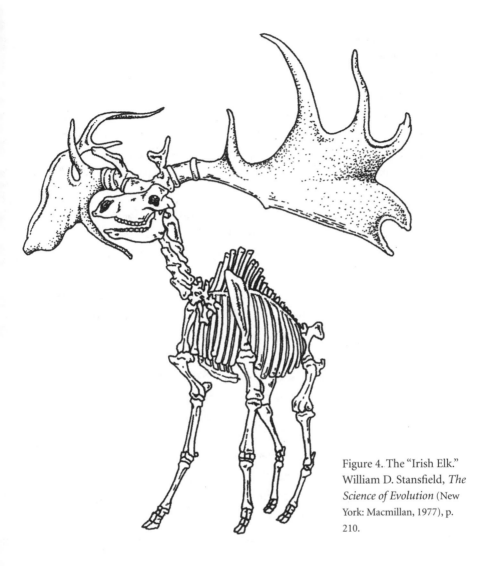

Figure 4. The "Irish Elk." William D. Stansfield, *The Science of Evolution* (New York: Macmillan, 1977), p. 210.

tage upon the "best adapted" individuals within a population. It would appear that, for a significant period of geological history, differential reproductive advantage was conferred instead upon increasingly maladapted individuals within the Irish elk population.

One may accommodate the data on *Megaloceros* either by making minor adjustments on the periphery of the theory or by changing the basic principles at the center. Some zoologists attacked the core principle of natural selection. Supporters of orthogenesis, for instance, took the history of *Megaloceros* to support the position that the direction of evolution is determined by immutable laws. This direction is supposed to be achieved independently of environmental changes and selective pressures. The increase of antler size in *Megaloceros* presumably implements an evolutionary law of development. By taking this position, supporters of orthogenesis reject the basic Darwinian principle that attributes the course of evolution to natural selective pressure operating on an antecedently given pool of variants.

A less drastic response to the *Megaloceros* problem was to take the growth of antlers to be a concomitant of the clearly adaptive growth of body size. As George Gaylord Simpson put it, "The Irish elk had the largest body of any of the deer in this group [European stags], and its antlers are just the size to be expected if the inherited relative growth pattern remained the same. The real trend, then, was for increase in total or body size. Antler size just tagged along, or raced ahead, in accordance with a growth pattern evidently adaptive in origin and adaptive in living stags."[6]

Some champions of the "tagged along" response appended a genetic rationale. Irish elk may have possessed genes that cause antlers to grow more rapidly than the rest of the body (positive allometry). This growth differential supposedly continues after sexual maturity has been reached. Individual elk with the largest antlers have an advantage in the competition to leave progeny, but are at a disadvantage subsequently. Since sexual selection has been an emphasis within evolutionary theory since its inception, the "tagged along" response restores agreement within the "field of force" of the theory without disturbing the basic principles at the center.

Stephen J. Gould proposed a resolution of the *Megaloceros* problem that restores agreement with natural selection theory through a reinterpretation of the evidence. According to Gould, the *Megaloceros* evidence is fully consistent with natural selection theory. The development of larger and larger antlers is an adaptive development. If this is correct, then the *Megaloceros* evidence does not require changes within the "field of force" of evolutionary theory.

Gould suggested that antler size functions as a signal that establishes reproductive rights without the need for combat. If this trait was adaptive before reforestation of the elk's domain and subsequently nonadaptive, then the *Megaloceros* data conforms to the requirements of natural selection theory.[7]

STEPHEN. J. GOULD ON THE IRISH ELK

The positive allometry of antlers among adult males is a fact. However, the standard interpretation of it is but one possible inference from empirical data—and it is one that I have come to doubt very much. This interpretation holds that the enormous antlers are a passive consequence of selection for larger bodies. Curiously, this standard contention has not escaped the dead hand of the orthogenetic explanation that it bravely claimed to replace. For it assumes that the antlers were disadvantageous *per se,* and that selection preserved them only because it favored the total phenotype of larger bodies and antlers. Instead of an immutable trend, we now have an immutable correlation.

Thus, the assumption of deleterious antlers was transported bodily from the orthogenetic to the allometric argument; the latter is, in fact, an attempt to preserve a Darwinian explanation within the assumption that antlers got too big. Yet the allometric relation prescribes no causes. It is equally subject to the opposite contention that selection acted primarily upon the antlers and engendered large bodies as a passive consequence. Or we may suppose (as I will argue presently) that larger bodies and relatively larger antlers were both favored by selection, and that the physiological correlation reinforced a rapid attainment of both conditions.

Why have students of *Megaloceros* been chained so long to an untested assumption that enormous antlers must be deleterious?

*   *   *

Now, if antlers are indeed weapons, then I do not see how we can avoid the assumption that they were developed beyond their optimum in *Megaloceros*. During the Alleröd, Ireland was free of large predators; wolves were probably the only natural enemy of *Megaloceros*. The interaction of wolves and moose indicates that large antlers are not needed for protection and that large body size itself suffices for adults in good condition.

*   *   *

If the antlers of *Megaloceros* were adapted primarily for display rather than for combat, then their enormous size can immediately be placed in a functional context. We know that relatively large antlers mark large bodies. Large bodied stags probably ranked highest in an order of dominance. If large antlers signalled this position without incurring the rigors of actual combat (either by female choice or, as I suspect, through ritualized encounters between competing males), then they would have been of the greatest possible and most immediate selective advantage—for they would have provided access to females and success in reproduction. The problem that led to orthogenetic proposals, and that rested on a hidden assumption that antlers must be weapons, simply vanishes.

* * *

When the short-lived Alleröd environment changed in Ireland—either to the last sharp cold spell of the Younger Dryas or to the luxuriant forestation that followed soon after as glaciers retreated throughout the northern hemisphere—*Megaloceros* became extinct. The antlers that had served it so well in open environments, may have seriously reduced its potential success in thick forests.

The case of the Irish elk illustrates the range of response options available to *prima facie* falsifying evidence.

3. Are There Crucial Experiments?

Scientists usually formulate *modus tollens* arguments in the service of some alternative hypothesis. Suppose H_1 and H_2 are in competition. If a sound *modus tollens* argument is formulated against H_2, then the standing of H_1 is enhanced. The strongest support would be achieved if it could be shown that one of the hypotheses H_1 and H_2 must be true, and that H_2 is false. An experiment that accomplished this would qualify as a "crucial experiment."

Francis Bacon spoke of such experiments as "crucial instances" or "instances of the fingerpost." A "fingerpost" is a road sign at an intersection that indicates a mutually exclusive choice of routes. Having arrived at such an intersection the scientist may ascertain which path to take by reference to a properly constructed experiment.

Suppose the two directions on a fingerpost are (1) that the free fall of bodies is due to attraction exerted by the mass of the earth, and (2) that the free fall of bodies is due to an intrinsic tendency of bodies to approach the center of the earth.

BACON ON INSTANCES OF THE FINGERPOST

The following therefore could be a Crucial Instance on this subject. Take a clock of the sort that works by leaden weights, and another that works by compression of an iron spring. They should be properly tested, so that neither is faster or slower than the other. Then have the clock working by weights placed at the top of some very high church and the other kept below, and let careful note be taken whether the clock placed on high goes more slowly than usual, on account of the diminished power of its weights. A similar experiment should be made in the depths of mines sunk deep beneath the earth, to see whether a clock of the same kind does not move faster than usual, through the increased

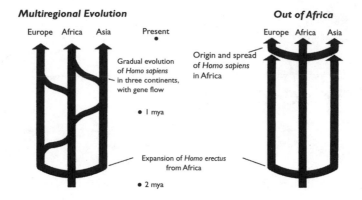

Multiregional Evolution

Europe Africa Asia   Present

Gradual evolution
of *Homo sapiens*
in three continents,
with gene flow

• 1 mya

Expansion of *Homo erectus*
from Africa

• 2 mya

Out of Africa

Europe Africa Asia

Origin and spread
of *Homo sapiens*
in Africa

Figure 5. Theories of the Origins of Modern Humans. From *The Origin of Humankind* by Richard Leakey, p. 87. Copyright © 1994 by B. V. Sherma. Reprinted by permission of Basic Books, a member of Perseus Book. L.L.C.

power of its weights. If the power of the weights is found to be diminished when the clock is high up and increased when it is underground, then we may take it that attraction by the bodily mass of the earth is the cause of weight.[1]

Bacon gave no indication that he had performed the crucial experiment he had described.

A recent candidate for "crucial" status is the "mitochondrial Eve" hypothesis. In *The Origin of Humankind*, Richard Leakey contrasted two theories about the origins of modern humans—multiregional evolution and the "out of Africa" theory. Leakey noted that recent evidence from molecular genetics may be a crucial experiment that eliminates the multiregional theory and enshrines the "out of Africa" theory.[2]

LEAKEY ON THE ORIGINS OF MODERN MAN

The third line of evidence bearing on the origin of modern humans, that of molecular genetics, is the least equivocal. It is also the most controversial. During the 1980s, a new model of modern human origins emerged. Known as the mitochondrial Eve hypothesis, it essentially supported the "Out of Africa" model, cogently so. Most proponents of the "Out of Africa" hypothesis are prepared to entertain the possibility that as modern humans expanded from Africa to the rest of the Old World they interbred to some degree with established premodern populations. This would allow for some threads of genetic continuity from ancient populations through to modern ones. The mitochondrial Eve model, however, refutes this. According to this model, as modern populations migrated out of Africa and grew in numbers, they completely replaced existing premodern populations. Interbreeding between the immigrant and existing populations, if it occurred at all, did so to an infinitesimal degree.

The mitochondrial Eve model flowed from the work of two laboratories—that of Douglas Wallace and his colleagues at Emory University, and of Allan Wilson and his colleagues at the University of California, Berkeley. They scrutinized the genetic material, or DNA, that occurs in tiny organelles within the cell called mitochondria. When an egg from a mother and sperm from a father fuse, the only mitochondria that become part of the cells of the newly formed embryo are from the egg. Therefore, mitochondrial DNA is inherited solely through the maternal line.

For several technical reasons, mitochondrial DNA is particularly suited to peering back through the generations in order to glimpse the course of evolution. And since the DNA is inherited through the maternal line, it eventually leads to a single ancestral female. According to the analyses, modern humans can trace their genetic ancestry to a female who lived in Africa perhaps 150,000 years ago. (It should be borne in mind, however, that this one female was part of a population of as many as 10,000 individuals; she was not a lone Eve with her Adam.)

Not only did the analyses indicate an African origin for modern humans, but they also revealed no evidence of interbreeding with premodern populations. All the samples of mitochondrial DNA analyzed so far from living human populations are remarkably similar to one another, indicating a common, recent origin. If genetic mixing between modern and archaic sapiens had occurred, some people would have mitochondrial DNA very different from the average, indicating its ancient origin. So far, with more than 4,000 people from around the world having been tested, no such ancient mitochondrial DNA has been found. All the mitochondrial DNA types from modern populations that have been examined appear to be of recent origin. The implication is that modern newcomers completely replaced ancient populations—the process having begun in Africa 150,000 years ago and then having spread through Eurasia over the next 100,000 years.

When Allan Wilson and his team first published their results, in a January 1987 issue of *Nature*, the conclusions were stated boldly, provoking consternation among anthropologists and wide interest among the public. Wilson and his colleagues wrote that their data indicated that "the transformation of archaic to modern forms of Homo sapiens occurred first in Africa, about 100,000 to 140,000 years ago, and . . . all present-day humans are descendants of that population." (Later analyses produced slightly earlier dates.) Douglas Wallace and his colleagues generally supported the Berkeley group's conclusions.

Milford Wolpoff stuck to his mulitregional model of evolution and denounced the data and analyses as unsound, but Wilson and his colleagues continued to produce more data and eventually stated that the conclusions were statistically unassailable. Recently, however, some statistical problems in the analysis were discovered, and the conclusions were recognized as being less concrete than had been asserted. Nevertheless, many molecular biologists still believe that the mitochondrial DNA data sufficiently support the "Out of Africa" hypothesis.

Of course, the mitochondrial Eve data is crucial only if (1) the out of Africa theory is confirmed, and (2) the out of Africa theory and the multiregional theory are the only possible theories of the origin of modern humans. The central difficulty for the notion of a "crucial experiment" is the requirement that the theories under consideration be shown to exhaust the set of possible interpretations.

Isaac Newton claimed in a letter to the Royal Society in 1671 that his two-prism experiment was an "*experimentum crucis.*"[3] He reasoned that if the hypothesis that sunlight is composed of rays of different colors with corresponding refractive properties is true, then passing a beam of light of a particular color through a prism results in deflection of the beam through the angle characteristic of that color, but no resolution of the beam into other colors. Newton passed a beam of sunlight through a prism in a darkened room, separated out, for example, the red portion of the beam and placed a second prism in the path of this red beam. He found that the second prism did deflect the red beam through the same angle as the angle that produced red in the case of the first prism. And there was no separation of the red beam into further colors by the second prism.[4]

Newton's prism experiments conform to a hypothetico-deductive model of scientific procedure. He performed the single-prism experiment,

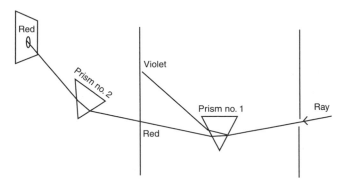

Figure 6. Newton's Two-Prism Experiment. The author's schematic representation of Newton's experiment recorded in *Opticks*, figure 18 facing p. 46 (New York: Dover, 1952).

induced the hypothesis that sunlight is composed of rays of different colors, each with a specific degree of refraction, deduced further consequences from the hypothesis—the two-prism experiment—and performed the appropriate test of these consequences.However, although the experimental confirmation achieved is striking, it remains to be shown that the two-prism experiment is a *crucial* experiment. To be such, the experimental result would have to disqualify competing hypotheses.

Newton's hypothesis is a conjunction of two claims: (1) each color is refracted by a prism through a characteristic angle, and (2) sunlight is composite, made up of the various spectral colors. The single-prism experiment provided the inductive basis for the first claim. Previous investigations had reported that the image produced by passing a beam of sunlight through a prism is roughly circular, with a violet fringe on the top (at the greatest angle of refraction) and a red fringe on the bottom. Newton found that, given a sufficient distance from prism to wall, a cylinder of spectral colors is formed.

The two-prism experiment provided additional evidential support for the correlation of color and degree of refraction, but there was no viable competing hypothesis about this correlation that was disqualified. If the two-prism experiment is "crucial," it is so with respect to the composition of sunlight hypothesis (the second conjunct).

Robert Hooke denied that Newton's experiments eliminated the obvious competing hypothesis. Hooke declared that "he doth not bring any argument to prove that all colours were actually in every ray of light before it suffered a refraction, nor does his *experimentum Crucis* as he calls it prove that those proprietys of coloured rayes, which we find they have after their first Refraction, were Not generated by the said Refraction."[5]

Hooke stressed that the observed data can be explained equally well by the hypothesis that a homogeneous beam of sunlight produces differently colored rays upon interaction with the glass of the prism. This competing hypothesis does not require that the red rays incident upon the second prism also interact with the glass to produce further colors. According to Hooke, the two-prism experiment establishes only that the same colors are associated with the same angles at each prism.

Newton did not use the phrase "*experimentum crucis*" in his description of the two-prism experiment in the *Opticks* (1704). However, he did introduce his discussion of the prism experiments under the heading "The Light of the Sun consists of Rays differently Refrangible—The Proof by Experiments."[6] Newton's claim to have achieved "proof" reflects his conviction, not shared by Hooke and other critics, that the composition of sunlight hypothesis is beyond doubt.

Duhem argued that no experiment can be crucial. Many mid-nineteenth-century scientists believed otherwise. Experiments performed by Fizeau and Foucault established that the velocity of light is greater in air than in water. This result is consistent with the Young-Fresnel wave theory but inconsistent with the Newtonian corpuscular theory. Many physicists took the Fizeau-Foucault result to be a crucial experiment that demonstrated that the corpuscular theory is false and the wave theory true. They held that light really is a wave motion and not a stream of particles. Duhem disagreed.[7]

DUHEM ON FOUCAULT'S EXPERIMENT TO "PROVE" THE CORPUSCULAR NATURE OF LIGHT

Suppose, for instance, we are confronted with only two hypotheses. Seek experimental conditions such that one of the hypotheses forecasts the production of one phenomenon and the other the production of quite a different effect; bring these conditions into existence and observe what happens; depending on whether you observe the first or the second of the predicted phenomena, you will condemn the second or the first hypothesis; the hypothesis not condemned will be henceforth indisputable; debate will be cut off, and a new truth will be acquired by science. Such is the experimental test that the author of the *Novum Organum* called the "fact of the cross, borrowing this expression from the crosses which at an intersection indicate the various roads."

We are confronted with two hypotheses concerning the nature of light; for Newton, Laplace, or Biot light consisted of projectiles hurled with extreme speed, but for Huygens, Young, or Fresnel light consisted of vibrations whose waves are propagated within an ether. These are the only two possible hypotheses as far as one can see; either the motion is carried away by the body it excites and remains attached to it, or else it passes from one body to another. Let us pursue the first hypothesis; it declares that light travels more quickly in water than in air; but if we follow the second, it declares that light travels more quickly in air than in water. Let us set up Foucault's apparatus; we set into motion the turning mirror; we see two luminous spots formed before us, one colorless, the other greenish. If the greenish band is to the left of the colorless one, it means that light travels faster in water than in air, and that the hypothesis of vibrating waves is false. If, on the contrary, the greenish band is to the right of the colorless one, that means that light travels faster in air than in water, and that the hypothesis of emissions is condemned. We look through the magnifying glass used to examine the two luminous spots, and we notice that the greenish spot is to the right of the colorless one; the debate is over; light is not a body, but a vibratory wave motion propagated by the ether; the emission hypothesis has had its day; the wave hypothesis has been put beyond doubt, and the crucial experiment has made it a new article of the scientific credo.

What we have said in the foregoing paragraph shows how mistaken we should be to attribute to Foucault's experiment so simple a meaning and so decisive an importance; for it is not between two hypotheses, the emission and wave hypotheses, that Foucault's experiment judges trenchantly; it decides rather between two sets of theories each of which has to be taken as a whole, i.e., between two entire systems, Newton's optics and Huygens' optics.

But let us admit for a moment that in each of these systems everything is compelled to be necessary by strict logic, except a single hypothesis; consequently, let us admit that the facts, in condemning one of the two systems, condemn once and for all the single doubtful assumption it contains. Does it follow that we can find in the "crucial experiment" an irrefutable procedure for transforming one of the two hypotheses before us into a demonstrated truth? Between two contradictory theorems of geometry there is no room for a third judgment; if one is false, the other is necessarily true. Do two hypotheses in physics ever constitute such a strict dilemma? Shall we ever dare to assert that no other hypothesis is imaginable? Light may be a swarm of projectiles, or it may be a vibratory motion whose waves are propagated in a medium; is it forbidden to be anything else at all? Arago undoubtedly thought so when he formulated this incisive alternative: Does light move more quickly in water than in air? "Light is a body. If the contrary is the case, then light is a wave." But it would be difficult for us to take such a decisive stand; Maxwell, in fact, showed that we might just as well attribute light to a periodical electrical disturbance that is propagated within a dielectric medium.

Unlike the reduction to absurdity employed by geometers, experimental contradiction does not have the power to transform a physical hypothesis into an indisputable truth; in order to confer this power on it, it would be necessary to enumerate completely the various hypotheses which may cover a determinate group of phenomena; but the physicist is never sure he has exhausted all the imaginable assumptions. The truth of a physical theory is not decided by heads or tails.

We know, in retrospect, that there is a third possibility, wave-particle dualism. In certain experimental contexts light manifests wavelike behavior. But in other experimental contexts, light manifests particle-like behavior. Scientists subsequently have been more cautious about claims to have exhausted the range of possible competing hypotheses. The only exclusive disjunction—one that rules out a third alternative—is (H_1 or $\sim H_1$). But Duhem had shown that falsification never is decisive against a single isolated hypothesis. Consequently $\sim H_1$ cannot be established as the victorious competitor by showing that H_1 is false.

John Herschel held, however, that an experiment can be "crucial" even though no proof is available that *no* other hypothesis is consistent with the result in question. He suggested that the first such experiment had been performed by Blaise Pascal in 1647.[8]

Pascal had become familiar with Torricelli's demonstration that if a glass tube is filled with mercury and raised to a vertical position in a pan containing that liquid, a 30-inch column of mercury remains in the tube.

Two hypotheses were proposed to account for this phenomenon. The first hypothesis is that there is a uniform force which resists the formation of a vacuum. The magnitude of this force is presumably just sufficient to support a 30-inch column of mercury (regardless of the diameter of the tube). The second hypothesis is that a "sea of air" presses down upon the surface of the mercury in the pan thereby supporting the weight of the mercury in the tube. This second hypothesis presupposes that pressure is transmitted undiminished throughout the fluid. Pascal recognized this. He took this presupposition to constitute a "law of hydrostatic pressure."

Pascal proposed a test to decide the issue between the hypothesis that nature abhors a vacuum and the sea of air hypothesis. He reasoned that if the sea of air hypothesis is correct, then the height of mercury supported at the top of a mountain is less than the height supported at its base. On the other hand, if the abhorrence of a vacuum hypothesis is correct, then there is no difference in height at the top and the base. Pascal instructed his brother-in-law to ascend the Puy de Dôme in central France with one of two similarly constructed barometers. Florin Perier reported that the height of the mercury column in the barometer carried to the top of the mountain was lower than the height of the mercury column in the barometer that remained at the base of the mountain.[9]

PASCAL'S PROOF OF THE SEA OF AIR HYPOTHESIS

First, I poured into a vessel six pounds of quicksilver which I had rectified during the three days preceding; and having taken glass tubes of the same size, each four feet long and hermetically sealed at one end but open at the other, I placed them in the same vessel and carried out with each of them the usual vacuum experiment. Then, having set them up side by side without lifting them out of the vessel, I found that the quicksilver left in each of them stood at the same level, which was twenty-six inches and three and a half lines above the surface of the quicksilver in the vessel. I repeated this experiment twice at this same spot, in the same tubes, with the same quicksilver, and in the same vessel; and found in each case that the quicksilver in the two tubes stood at the same horizontal level, and at the same height as in the first trial.

That done, I fixed one of the tubes permanently in its vessel for continuous experiment. I marked on the glass the height of the quicksilver, and leaving that tube where it stood, I requested Revd. Father Chastin, one of the brothers of the house, a man as pious as he is capable, and one who reasons very well upon these matters, to be so good as to observe from time to time all day any changes that might occur. With the other tube and a portion of the same quicksilver, I then proceeded with all these gentlemen to the top of the Puy de Dôme, some 500 fathoms above the Convent. There, after I had made the same experiments in the same way that I had made them at the Minims, we found that there remained in the tube a height of only twenty-three inches and two lines of quicksilver; whereas in the same tube, at the Minims we had found a height of twenty-six inches and three and a half lines. Thus between the heights of the quicksilver in the two experiments there proved to be a difference of three inches one line and a half. We were so carried away with wonder and delight, and our surprise was so great that we wished, for our own satisfaction, to repeat the experiment. So I carried it out with the greatest care five times more at different points on the summit of the mountain, once in the shelter of the little chapel that stands there, once in the open, once shielded from the wind, once in the wind, once in fine weather, once in the rain and fog which visited us occasionally. Each time I most carefully rid the tube of air and in all these experiments we invariably found the same height of quicksilver. This was twenty-three inches and two lines, which yields the same discrepancy of three inches, one line and a half in comparison with the twenty-six inches, three lines and a half which had been found at the Minims. This satisfied us fully.

Of course, the Puy de Dôme experiment is not crucial in the sense that only the sea of air hypothesis is consistent with the result. The abhorrence of a vacuum hypothesis can be modified to account for the decrease in level with height. All that is required is to make the force that resists the formation of a vacuum dependent on altitude—the less dense the air, the less powerful the force of abhorrence. This modification may be rejected because it appears to be an arbitrary attempt to evade falsification, but the Puy de Dôme result itself does not falsify it. The Puy de Dôme result does not even prove false the hypothesis that there is a constant force of abhorrence throughout the universe. The abhorrence theorist can retain this hypothesis but add that there exists a second, opposing force that becomes stronger with altitude. What is falsified by the experiment is the hypothesis that mercury levels are sustained by one and only one force whose magnitude is constant at all locations.

Relatively crucial experiments also have been advanced in evolutionary biology. If paleontological evidence is correctly interpreted, there was a time in the earth's history before which no insects possessed wings. But if wingless insects evolved into winged insects capable of flight, then there is a "how possibly" puzzle. How could insects come to develop wings that sustain flight, given that any "protowings" that arise initially would be insufficiently strong to enable flight? Protowings would confer no adaptive advantage in this respect. To resolve the puzzle one needs to formulate a "how possibly" explanation. Robert Brandon notes that "how-possibly explanations are potential explanations, none of whose explanatory premises contradict or conflict with 'known facts.'"[10] Such explanations are common in evolutionary biology.

Joel Kingsolver and M. A. R. Koehl entertained two hypotheses to solve the above puzzle:[11]

H_1: Protowings do provide adaptive advantage because (1) they enable insects to glide, or (2) they slow insects' rate of fall to the earth, or (3) they improve the stability of insects' landings.

H_2: Protowings provide adaptive advantage because they enable insects to improve the regulation of body temperature.

Kingsolver and Koehl performed experiments on artificial model insects to decide the issue between the aerodynamic hypothesis (H_1) and the thermoregulation hypothesis (H_2). In one set of experiments, model insects with protowings whose lengths are less than 30% of the model insect body length displayed no aerodynamic benefits. Insects with protowings fared no better gliding and landing in the wind tunnel than did insects without protowings.

The wind tunnel experiments are not "crucial" experiments that falsify H_1 and enthrone H_2, even if it is conceded that the behavior of model insects faithfully reproduces the behavior of real insects. H_1 and H_2 are not the only possible hypotheses to account for the emergence of winged insects. Kingsolver and Koehl themselves conceded that their experimental results do not count against hypotheses that take moveable protowings to have evolved from the gills of aquatic insect species to facilitate ventilation or locomotion.[12] In addition, it is possible, although highly unlikely on the basis of current knowledge of genetics, that macromutations to full-sized wings occurred at a point in time and subsequently became dominant in the insect population. The macromutation hypothesis also is consistent with the wind tunnel results.

Nevertheless, the wind tunnel experiments count against aerodynamic hypothesis H_1 and thereby indirectly support thermoregulation hypothesis H_2. Kingsolver and Koehl also provided more direct evidence for H_2. They showed that model insects with protowings achieved superior regulation of body temperature when winged and wingless model insects were exposed to intermittent radiation from a flood lamp.

Kingsolver and Koehl showed, in addition, that there is a ratio of wing size to body length above which aerodynamic advantages begin and thermoregulatory advantages decrease. If these results for model insects are applicable to their real-life counterparts, then it is likely that the evolution of winged insects is a two-stage process. The first stage is the emergence of protowinged insects that possess a thermoregulatory advantage over competitors, and the second stage is the emergence of full-winged insects that possess an aerodynamic advantage over competitors.

KINGSOLVER AND KOEHL ON THE EVOLUTION OF INSECT WINGS

The purpose of this study is to evaluate several aerodynamic and thermoregulatory hypotheses about the evolution of wings. Our approach is to construct a series of physical models of Paleozoic insects, and to measure aerodynamic and heat transfer characteristics of these models as functions of body size and shape, and the number, size, and physical characteristics of wings. Our results illustrate how the differences by which aerodynamics and heat transfer scale with body and wing size have important consequences for the evolution of insect wings.

＊　＊　＊

One can identify three distinct roles for wings in relation to aerodynamics: gliding, parachuting or ballooning, and attitude stability or control. Note that the gliding and parachuting hypotheses may apply to either fixed or moveable wings, while the attitude stability hypothesis assumes moveable wings.

＊　＊　＊

The Models

Models were made of insects with two different body shapes, representing both flying and non-flying forms; these were based on published reconstructions of Paleozoic insect fossils.

* * *

Based on morphological and systematic evidence, it is generally believed that winged insects arose from ancestors of about 2–4 cm body length during the early to late Devonian (Wootton, 1976; Wigglesworth, 1976). Given these uncertainties, any understanding of selective factors operating during this crucial period in insect evolution must be speculative. At best, we can eliminate certain hypotheses as untenable and document other hypotheses as at least plausible.

Our results indicate that, of the three aerodynamic hypotheses that have been proposed for selection for insect wings, there are no significant aerodynamic differences between wingless and shortwinged insects. One possible exception is for relative maximum lift/drag (i.e., glide distance) for the 10 cm model, for which wings less than 1 cm long may have important effects. For insects in the size range from which winged insects probably arose, wing lengths more than 30–60% of body length are required before there are any significant aerodynamic effects of wings. Thus, gradual selection for increased wing length in wingless insects could not occur on the basis of aerodynamic performance.

* * *

On the other hand, as suggested by Douglas (1981), short wings can have large thermoregulatory effects, particularly at the small body sizes probably typical of the early insects. For these small sizes, increasing wing length significantly increases temperature excess for wing lengths up to 20–40% of body length. These effects are largely independent of details of body and wing shape, radia-

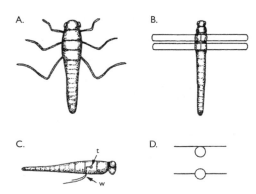

Figure 7. Model Insects. Joel Kingsolver and M. A. R. Koehl, "Aerodynamics, Thermoregulation, and the Evolution of Insect Wings: Differential Scaling and Evolutionary Change," *Evolution* *39* (1985), p. 491. Permission courtesy of the Society for the Study of Evolution.

tion and wind conditions, and the conductance of heat through the wings. Our transient analyses indicate that, for wings of low thermal conductivity (as in present-day insects), the wings act to increase the effective surface area for radiation absorption, rather than to decrease the heat transfer coefficient. For wings of high thermal conductivity, wings both greatly increase the effective surface area and increase the heat transfer coefficient.

* * *

Our results are consistent with the hypothesis that the initial evolution of wings from ancestors with small winglets was related to selection for increased thermoregulatory capacity, which would be particularly effective at the small body sizes of the earliest insects. After this initial period, effective selection for increased aerodynamic capacity could occur. For small insects, this could only occur for wing lengths greater than about 50–60% of body length; for larger insects this could occur at relatively smaller wing lengths. Thus, we propose that thermoregulation was the primary adaptive factor in the early evolution of wings, preadapting them for the subsequent evolution of flight.

* * *

We have shown that there is a switch in functional capacity from thermoregulation to aerodynamics with increasing wing length at all body sizes. However, the relative wing length, or geometry, at which this switch occurs depends on body size. This means that geometrically identical (i.e., isomorphic) forms may serve different functions at different body sizes. Consider, for example, an insect with wings 50% as long as the body. In a small insect, these wings could function effectively for thermoregulation, but not for aerodynamics; in a large insect, these same wings could serve quite effectively as aerodynamic structures. Thus, a purely isometric change in body size during evolution may yield a change in function of a given structure.

* * *

These considerations suggest an alternative scenario for the transition from a thermoregulatory to an aerodynamic function for insect wings. Elongation of the wings first evolved in small insects as a result of selection for thermoregulatory capacity, followed by an isometric increase—either gradual or abrupt—in body size, after which wings could function as aerodynamic structures. Thus, we argue that changes in body form were not a prerequisite for this major change in function.[13]

Despite the claims of Bacon and Newton, there are no crucial experiments in science. When scientists such as Herschel, or Kingsolver and Koehl, talk about (relatively) crucial experiments, there always is a caveat; the experiment is "crucial" between H_1 and H_2 only on the assumption that no third hypothesis is relevant.

4. Falsification and the Method of Difference

John Stuart Mill recommended a method of difference as a rule of proof of causal connectedness. He maintained that if it can be shown that the following schema is instantiated, where A, B, and C are the only possible causes of p, then A is a necessary condition of the occurrence of phenomenon p.

instance	circumstances	phenomenon
1	$A\,B\,C$	p
2	$B\,C$	———

If A is not present then p does not occur. Mill was willing to generalize that the pattern displayed in the two instances above will be realized for every pair of similar instances that involve circumstances of types A, B, and C (and no other circumstances). He noted that "the Method of Difference has for its foundation, that whatever cannot be eliminated is connected with the phenomenon by law."[1]

Mill overestimated what can be achieved by instantiation of the schema. The principal limitation of the method is our inability to list every circumstance that is different in the two instances under consideration. We may believe that the only circumstances relevant to the occurrence of p are A, B, and C. But in practice, two instances differ in innumerable respects—spatial location, time, the relative positions of the bodies in the solar system, the orientation of the observers in the laboratory, sunspot activity, cloud formations, and so on. Success in applying the method depends on whether or not circumstances not listed in the schema are causally relevant to the phenomenon under investigation. The method of difference also can be applied to falsify hypotheses. Suppose it is hypothesized that *C* is a necessary condition of the occurrence of *p*. If the following pattern is observed then that hypothesis is discredited.

instance	circumstances	phenomenon
1	*A B C*	*p*
2	*A B*	*p*

C cannot be a necessary condition of *p* if *p* occurs in its absence.

An example from the history of science is the discrediting of the aether hypothesis by the null result of the Michelson-Morley experiment. On the aether hypothesis, light waves are propagated in an invisible fluid medium. Aether theorists assumed that light waves have a specific definite velocity with respect to this medium. If this is the case, then the velocity of light as measured on earth should depend on the velocity of the earth with respect to the aether. The mathematics of the situation is such that the round-trip velocity of light parallel to the earth's motion through the aether should be less than the velocity of light perpendicular to the earth's motion through the aether.

Michelson and Morley constructed a device to compare the round-trip velocities of light parallel to, and perpendicular to, the earth's orbital motion. No significant difference in the two velocities was found.

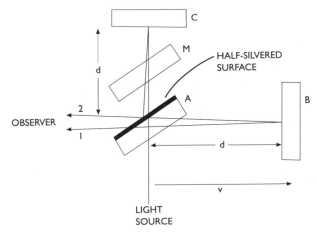

Figure 8. The Michelson-Morley Experiment. F. W. van Name, *Modern Physics* (New York: Prentice-Hall, 1952), p. 60.

EINSTEIN ON THE MICHELSON-MORLEY EXPERIMENT

For a long time the efforts of physicists were devoted to attempts to detect the existence of an aether-drift at the earth's surface. In one of the most notable of these attempts Michelson devised a method which appears as though it must be decisive. Imagine two mirrors so arranged on a rigid body that the reflecting surfaces face each other. A ray of light requires a perfectly definite time T to pass from one mirror to the other and back again, if the whole system be at rest with respect to the aether. It is found by calculation, however, that a slightly *different* time T' is required for this process, if the body, together with the mirrors, be moving relatively to the aether. And yet another point: it is shown by calculation that for a given velocity v with reference to the aether, this time T' is different when the body is moving perpendicularly to the planes of the mirrors from that resulting when the motion is parallel to these planes. Although the estimated difference between these two times is exceedingly small, Michelson and Morley performed an experiment involving interference in which this difference should have been clearly detectable. But the experiment gave a negative result—a fact very perplexing to physicists. Lorentz and FitzGerald rescued the theory from this difficulty by assuming that the motion of the body relative to the aether produces a contraction of the body in the direction of motion, the amount of contraction being just sufficient to compensate for the difference in time mentioned above. . . . also from the standpoint of the theory of relativity this solution of the difficulty was the right one. But on the basis of the theory of relativity the method of interpretation is incomparably more satisfactory. According to this theory there is no such thing as a "specially favoured" (unique) co-ordinate system to occasion the introduction of the aether-idea, and hence there can be no aether-drift, nor any experiment with which to demonstrate it. Here the contraction of moving bodies follows from the two fundamental principles of the theory, without the introduction of particular hypotheses; and as the prime factor involved in this contraction we find, not the motion in itself, to which we cannot attach any meaning, but the motion with respect to the body of reference chosen in the particular case in point. Thus for a coordinate system moving with the earth the mirror system of Michelson and Morley is not shortened, but it is shortened for a co-ordinate system which is at rest relatively to the sun.[2]

What aether theorists had expected to find was the following, where A is the circumstance that the direction of propagation of light and the earth's orbital motion are parallel, B and C are properties of the experimental apparatus, and d is a decrease in the round-trip velocity of light c in the direction of the earth's orbital motion.

instance	circumstances	phenomena
1	A B C	c, d
2	B C	c

Michelson and Morley, however, found the following result:

instance	circumstances	phenomena
1	A B C	c
2	B C	c

Many physicists took this result to indicate that the propagation of light waves does not involve a material medium.

Lorentz and Fitzgerald, by contrast, developed an interpretation of the null result that retained the "aether drift" that retards the round-trip velocity of light in the direction parallel to the earth's motion.[3] They suggested that this retarding effect might be canceled by a contraction of the measuring instruments in the direction of the earth's motion through the aether—circumstance D in the schema below, where $c^* = c - d + d$:

instance	circumstances	phenomenon
1	A B C D	c*
2	B C	c

The Lorentz-Fitzgerald interpretation, which preserves the role of the aether, also is consistent with the Michelson-Morley result. Instantiation of the difference schema does not, in itself, establish falsification. It always remains possible that there are relevant circumstances that have not been taken into account.

Mill was aware of the problem posed by relevant circumstances not taken into account. He discussed the case of arsenic poisoning. The correlation between ingesting arsenic and death receives support from numerous applications of the difference schema, for example, below, where A is the ingestion of (more than a specific amount of) arsenious acid, d is death and B, C, and D are other circumstances presumably the same for the two instances.

instance	circumstances	phenomenon
1	A B C D	d
2	B C D	——

If *A*, *B*, *C*, and *D* are the only relevant circumstances, then *A* is the necessary condition of particular death *d*.

Mill pointed out that the generalization that *A* causes *d* regardless of the other circumstances present is falsified by instances in which an antidote is present. Given the arsenic-death generalization, we anticipate that, regardless of circumstance *X*, the following will be observed:

instance	circumstances	phenomenon
1	A B C D X	d
2	A B C D	d

However, Mill noted that if *X* is the prior administration of an antidote such as hydrated peroxide of iron, then what is observed is the following:

instance	circumstances	phenomenon
1	A B C D X	——
2	A B C D	d

Mill concluded that it is false that ingestion of arsenious acid is invariably and unconditionally associated with subsequent death.[4]

Mill emphasized that circumstances can reinforce or cancel one another. The poison-antidote relation is an illustration of cancellation. (The Lorentz-Fitzgerald interpretation of the Michelson-Morley experiment also invokes a cancellation relation.) Reinforcement and cancellation are widespread in physics. Mill called attention to the composition of forces.[5] The resultant force F_R of two component forces F_1 and F_2 is the diagonal of a parallelogram:

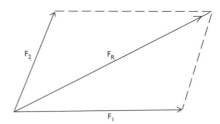

Figure 9. The Parallelogram of Forces.

Since innumerably many component-force combinations can produce resultant force F_R, one cannot infer the "component-force causes" operating from information about F_R alone. The method of difference is of no help in situations in which multiple causes produce effects that differ from the effects produced by each cause separately.

Despite its limitations, the method of difference is of great value in the context of scientific discovery. The difference schema is an ideal that scientists seek to realize by setting up controlled experiments. Given that there are cases for which the relevant circumstances can be controlled, applications of the method of difference may succeed in identifying causal relations.

Mill went further. He maintained that the difference schema is a rule of proof of causal connections:

instance	laws that are possible p-explainers	explanandum
1	L_1, L_2, L_3	p
2	L_2, L_3	_____

$$\therefore L_1 \text{ is the explanation of } p$$

If it can be shown that there is a correct explanation of p, that L_1, L_2, and L_3 are the only possible p-explainers, and that L_2 and L_3 fail to explain p, then L_1 is the correct explanatory law. Laws gain certification within the history of science by satisfying this schema. However, he gave just one example—Newton's explanation why a planet moves in a closed orbit subject to Kepler's law of areas. (This law states that the radius vector from the sun to a planet sweeps over equal areas in equal times.) According to Newton, the planet obeys Kepler's law because of a $1/r^2$ central force emanating from the sun. He showed that if $n = 2$ then the planet has a stable orbit and obeys the law of areas, but that if n has any value other than 2 then the planet has no closed orbit at all. It was not necessary for Newton to make calculations for every integral and decimal value of n. He was able to prove that as the value of n increases or decreases from 2, there is increasing divergence from Keplerian ellipses and the law of areas.

MILL'S "PROOF" OF THE LAW OF GRAVITATIONAL ATTRACTION

We want to be assured that the law we have hypothetically assumed is a true one; and its leading deductively to true results will afford this assurance, provided the case be such that a false law cannot lead to a true result; provided no law, except the very one which we have assumed, can lead deductively to the same conclusions which that leads to. And this proviso is often realized. For example, in the very complete specimen of deduction which we just cited, the original major premise of the ratiocination, the law of the attractive force, was ascertained in this mode; by this legitimate employment of the Hypothetical Method. Newton began by an assumption, that the force which at each instant deflects a planet from its rectilineal course, and makes it describe a curve round the sun, is a force tending directly towards the sun. He then proved that if this be so, the planet will describe, as we know by Kepler's first law that it does describe, equal areas in equal times; and, lastly, he proved that if the force acted in any other direction whatever, the planet would not describe equal areas in equal times. It being thus shown that no other hypothesis would accord with the facts, the assumption was proved; the hypothesis became an inductive truth. Not only did Newton ascertain by this hypothetical process the direction of the deflecting force; he proceeded in exactly the same manner to ascertain the law of variation of the quantity of that force. He assumed that the force varied inversely as the square of the distance; showed that from this assumption the remaining two of Kepler's laws might be deduced; and finally, that any other law of variation would give results inconsistent with those laws, and inconsistent, therefore, with the real motions of the planets, of which Kepler's laws were known to be a correct expression.

I have said that in this case the verification fulfils the conditions of an induction: but an induction of what sort? On examination we find that it conforms to the canon of the Method of Difference. It affords the two instances, A B C, a b c, and B C, b c. A represents central force; A B C, the planets *plus* a central force; B C, the planets apart from a central force. The planets with a central force give a, areas proportional to the times; the planets without a central force give b c (a set of motions) without a, or with something else instead of a. This is the Method of Difference in all its strictness. It is true, the two instances which the method requires are obtained in this case, not by experiment, but by a prior deduction. But that is of no consequence. It is immaterial what is the nature of the evidence from which we derive the assurance that A B C will produce a b c, and B C only b c; it is enough that we have that assurance. In the present case, a process of reasoning furnished Newton with the very instances, which, if the nature of the case had admitted of it, he would have sought by experiment.

It is thus perfectly possible, and indeed is a very common occurrence, that what was a hypothesis at the beginning of the inquiry, becomes a proved law of nature before its close. But in order that this should happen, we must be able, either by deduction or experiment, to obtain both the instances which the Method of Difference requires. That we are able from the hypothesis to deduce the known facts, gives only the affirmative instance, *A B C, a b c*. It is equally necessary that we should be able to obtain, as Newton did, the negative instance *B C, b c*; by showing that no antecedent, except the one assumed in the hypothesis, would in conjunction with *B C* produce *a*.

Now it appears to me that this assurance cannot be obtained, when the cause assumed in the hypothesis is an unknown cause, imagined solely to account for *a*. When we are only seeking to determine the precise law of a cause already ascertained, or to distinguish the particular agent which is in fact the cause, among several agents of the same, one or other of which it is already known to be, we may then obtain the negative instance. An inquiry, which of the bodies of the solar system causes by its attraction some particular irregularity in the orbit or periodic time of some satellite or comet, would be a case of the second description. Newton's was a case of the first. If it had not been previously known that the planets were hindered from moving in straight lines by some force tending towards the interior of their orbit, though the exact direction was doubtful; or if it had not been known that the force increased in some proportion or other as the distance diminished, and diminished as it increased; Newton's argument would not have proved his conclusion. These facts, however, being already certain, the range of admissible suppositions was limited to the various possible directions of a line, and the various possible numerical relations between the variations of the distance, and the variations of the attractive force: now among these it was easily shown that different suppositions could not lead to identical consequences.[6]

Mill took this explanation to fulfill the requirements of the difference schema which eliminates all possible explanations except one. Mill's analysis of Newton's achievement overlooks the fact that Kepler's laws hold only approximately. What Newton explained in the argument discussed by Mill is not the motion of a planet but the motion of a point-mass in the $1/r^2$ gravitational field of a stationary point-center of force. In the real world the gravitational force between sun and planet is reciprocal, and the presence of other planets in the solar system results in perturbations of a planet's orbit.

Newton was well aware of this. Indeed, he made the distinction between the ideal realm of mathematical relations and the real world of interacting physical bodies a foundation of his physics. Newton attributed the small deviations from Kepler's laws to reciprocal gravitational forces among the bodies of the solar system. Subsequently, Laplace demonstrated that the perturbations caused by the reciprocal attractions of Jupiter and Saturn are cyclical, such that the sun-Jupiter-Saturn system is dynamically stable.[7]

Mill's claim that an instantiation of the difference schema *proves* the law of gravitational attraction must be rejected. In general, no instantiation of the difference schema could establish "verification" in Mill's sense of the term. Just as it cannot be proved that a set of circumstances includes every relevant circumstance, it cannot be proved that a set of explanations includes every possible explanation. Applications of the method of difference, like experiments deemed to be "crucial," achieve "falsification" only subject to an unproven assumption. In the case of the difference schema, the assumption restricts the set of relevant circumstances to those explicitly stated in its instantiation.

2. THE REJECTION OF THEORIES

5. Popper's Methodological
   Falsificationism

Perhaps the appropriate response to our inability to prove scientific hypotheses false is to adopt the search for falsification as a criterion of empirical methodology. This is a position championed by Karl Popper. Popper maintained that the proper approach for a scientist to take is to become an antagonist against her own (and others') hypotheses. One practices proper empirical methodology by performing tests designed to refute proposed hypotheses.

According to Popper, it is the search for falsification that distinguishes empirical science from nonempirical methodologies. However, since attempted falsifications never are decisive, the demarcation achieved is fuzzy. There is no sharp circumference that separates empirical theories from nonempirical theories. Rather, there is a substantial donut-shaped region within which the empirical status of methodological practices is subject to dispute.

Methodological falsificationism requires that a scientist seek impor-

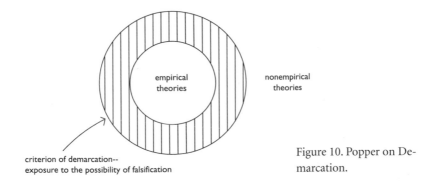

empirical
theories

nonempirical
theories

criterion of demarcation--
exposure to the possibility of falsification

Figure 10. Popper on Demarcation.

tant tests of her theories. However, it does not require her to reject a theory upon receipt of a single negative test result. The confrontation with negative evidence is never so compelling as to force one either to accept the falsehood of a theory or to be convicted of irrationality.

One reason that falsification never is decisive is that the observation report that allegedly falsifies a hypothesis may be false. Consider the generalization that "all blue litmus paper turns red in acid solution." An observer performs a test and reports that the paper is still blue. One can pose a number of questions about the observation report:

Is there a coating on the paper that insulates it from the liquid?
Is the paper not in the liquid at all?
Is the report about a holographic illusion rather than paper?
Is the observer red-blue color-blind?

If the answer to any one of such questions is yes, then the observation report is false and the attempt to falsify the generalization fails.

Popper conceded that the truth of an observation report is always subject to challenge. Observation reports are not incorrigible. Nevertheless, if the testing of theories is to be possible, an observational test basis must be accepted. Popper referred to the observation reports that constitute this test basis as "basic statements." Basic statements assert the existence of an event (object, property, relation) within a specific region of space and time. Such statements record the results of tests designed to maximize the likelihood that some law or theory will be found to be false.

POPPER ON "BASIC STATEMENTS"

From a logical point of view, the testing of a theory depends upon basic statements whose acceptance or rejection, in its turn, depends upon our *decisions*. Thus it is *decisions* which settle the fate of theories. To this extent my answer to the question, "how do we select a theory?" resembles that given by the conventionalist; and like him I say that this choice is in part determined by considerations of utility. But in spite of this, there is a vast difference between my views and his. For I hold that what characterizes the empirical method is just this: that the convention or decision does not immediately determine our acceptance of *universal* statements but that, on the contrary, it enters into our acceptance of the *singular* statements—that, is the basic statements.

For the conventionalist, the acceptance of universal statements is governed by his principle of *simplicity:* he selects that system which is the simplest. I, by contrast, propose that the first thing to be taken into account should be the severity of tests. (There is a close connection between what I call "simplicity" and the severity of tests; yet my idea of simplicity differs widely from that of the conventionalist. . . .) And I hold that what ultimately decides the fate of a theory is the result of a test, i.e. an agreement about basic statements. With the conventionalist I hold that the choice of any particular theory is an act, a practical matter. But for me the choice is decisively influenced by the application of the theory and the acceptance of the basic statements in connection with this application; whereas for the conventionalist, aesthetic motives are decisive.

Thus I differ from the conventionalist in holding that the statements decided by agreement are *not universal but singular.*[1]

Popper acknowledged that his falsificationalist methodology incorporates an element of conventionalism. It is a matter of convention that a scientist decides to accept a basic statement as true without asking further questions. It is necessary to accept as true some basic statement in order to test a hypothesis. The scientist accepts a basic statement knowing that the relevant state of affairs is incompletely (and perhaps inaccurately) described. No basic statement can specify every aspect of the state of affairs described. The scientist realizes that some aspects not described (or perhaps misdescribed) may be important to the test status of the basic statement. Nevertheless, the scientist accepts the statement in order to fulfill the methodological requirement that hypotheses be subjected to tests.

6. Responses to *Prima Facie* Falsifying Evidence

If an accepted basic statement counts against a theory, the theory may be rejected outright. But other options are available.

CONFRONTATION DEFERRED

The basic statement may be accepted as true, but shelved for future consideration. Mendeleyev took this approach when confronted with anomalous data on the relative atomic weights of tellurium and iodine. Mendeleyev had developed an arrangement of chemical elements on the basis of a periodic law which states that "the properties of simple bodies [elements], the constituents of their compounds, as well as the properties of these last, are periodic functions of the atomic weights of elements."[1]

In Mendeleyev's arrangement, vertical groups of elements form compounds of similar composition. For the lightest elements, the periodicity conforms to a law of octaves. (The law of octaves was stated by J. R. Newlands in 1865.)

Li = 7	Be = 9	B = 11	C = 12	N = 14	O = 16	F = 19
Na = 23	Mg = 24	Al = 27	Si = 28	P = 31	S = 32	Cl = 35.5

The formulas and properties of the compounds of Li and Na are similar, as are the formulas and properties of the compounds of F and Cl.

Mendeleyev realized that the requirement of periodicity places iodine beneath fluorine, chlorine, and bromine. However, the best available data on atomic weights revealed that the atomic weight of iodine (126) is less than that of tellurium (128). If elements were arranged in order of increasing atomic weight, then the grouping would be the following:

$$
\begin{array}{ll}
O = 16 & F = 19 \\
S = 32 & Cl = 35.5 \\
Se = 78 & Br = 80 \\
I = 126 & Te = 128
\end{array}
$$

Mendeleyev was unwilling to accept the iodine-tellurium weight relation as a falsification of the periodic law. He conceded that the atomic-weight determinations constitute an anomaly to the law but held out the prospect that further inquiry may provide a resolution.

MENDELEYEV ON THE TELLURIUM-IODINE ANOMALY

According to the periodic law, the atomic weight of tellurium should be greater than that of Sb = 122, and less than that of I = 127; that is to say that the atomic weight of tellurium ought to be about 125, because from every point of view, atomic analogies assign it a place between Sb and I.

* * *

It is difficult to purify the compounds of tellurium and even to be certain when they are pure. This may perhaps explain in a measure the errors of the numbers which have been found. It is difficult to admit that the distinctive individual characteristics of tellurium could determine a gap relatively so great (128 to 125) compared with the number of its atomic weight, as it is deduced from the periodic law. Fresh experiments are therefore necessary to show us to what degree the periodic law can be relied upon in the correction of atomic weights.[2]

Mendeleyev's position was not irrational. He had good reasons to shelve the iodine-tellurium anomaly. The periodic law was a principle that organized a vast amount of data on chemical elements and their compounds. Moreover, by disregarding the potential falsifying evidence of the iodine-tellurium inversion, he was able to apply the periodic law to make predictions about the properties of yet-to-be-discovered elements. Having arranged the known elements to conform to the periodic law (but with the positions of iodine and tellurium interchanged), Mendeleyev noted that there were gaps in the table.[3]

Mendeleyev highlighted the gaps below boron, aluminum, and silicon in his periodic arrangement. Knowledge of the variation of properties of adjacent elements and their compounds, in both the horizontal and vertical dimensions, enabled him to make detailed predictions about the yet-

2. Mendeleyev's Table of Elements (1871). J. W. van Spronsen, *The Periodic System of Chemical Elements* (New York: Elsevier, 1969), p. 137

	Gruppe I R^2O	Gruppe II RO	Gruppe III R^2O^3	Gruppe IV RO^2	Gruppe V RH^4 R^2O^5	Gruppe VI RH^3 RO^3	Gruppe VII RH^2 R^2O^7	Gruppe VIII RH RO^4
1	H=1							
2	Li=7	Be=9.4	B=11	C=12	N=14	O=16	F=19	
3	Na=23	Mg=24	A1=27.3	Si=28	P=31	S=32	C1=35.5	
4	K=39	Ca=40	−=44	Ti=48	V=51	Cr=52	Mn=55	Fe=56, Co=59 Ni=59, Cu=63
5	(Cu=63)	Zn=65	−=68	−=72	As=75	Se=78	Br=80	
6	Rb=85	Sr=87	?Yt=88	Zr=90	Nb=94	Mo=96	−=100	Ru=104, Rh=104 Pd=106, Ag=108
7	(Ag=108)	Cd=112	In=113	Sn=118	Sb=122	Te=125	J=127	
8	Cs=133	Ba=137	?Di=138	?Ce=140				
9	−							
10			?Er=178	?La=180	Ta=182	W=184		Os=195, Ir=197 Pt=198, Au=199
11	(Au=199)	Hg=200	Ti=204	Pb=207	Bi=208			
12				Th=231		U=240		

to-be-discovered elements.[4] These predictions were confirmed upon discovery of "eka-boron" (scandium) in 1879, "eka-aluminum" (gallium) in 1875 and "eka-silicon" (germanium) in 1886.

Of course, Mendeleyev's periodic law is false. The chemical properties of elements are periodic functions of their atomic numbers and not their atomic weights. The atomic number of an element is the total positive charge of its nucleus (the number of protons in the nucleus). On the basis of atomic number, the iodine-tellurium relation is no longer an anomaly. Tellurium has atomic number 52 and iodine has atomic number 53. That the atomic weight of tellurium is greater than that of iodine is the result of greater numbers of neutrons in the naturally occurring isotopes of tellurium. The most abundant isotopes of tellurium—$_{52}Te^{130}$ and $_{52}Te^{128}$—

have 78 and 76 neutrons in their nuclei, whereas naturally occurring iodine—$_{53}I^{127}$—possesses 74 neutrons.

Mendeleyev's periodic law is almost correct. Increasing atomic number is nearly always accompanied by increasing atomic weight. Mendeleyev, of course, was unaware of this. His decision to apply the (strictly false) periodic law proved fruitful. Had he been a strict falsificationist unwilling to put aside the iodine-tellurium anomaly, he would have passed up an important opportunity for scientific progress. The conversion of the

3. Mendeleyev's Predictions (1871). J. W. van Spronsen, *The Periodic System of Chemical Elements* (New York: Elsevier, 1969), p. 139

	Predictions	*Determinations*
	Eka*-aluminium	Gallium (discovered in 1875 by Lecoq de Boisbaudran)
at. w.	68	69.9
sp. w.	6.0	5.96
at. vol.	11.5	11.7
	Ekaboron	Scandium (discovered in 1879 by Nilson)
at. w.	44	43.79
oxide	Eb_2O_3 sp. w. 3.5	Sc_2O_3 sp. w. 3.864
sulphate	$Eb_2O(SO_4)_3$	$Sc_2(SO_4)_3$
bisulphate	not isomorphous with alum	small narrow columns
	Ekasilicon	Germanium (discovered in 1886 by Winkler)
at. w.	72	72.3
sp. w.	5.5	5.469
at. vol.	13	13.2
oxide	EsO_2	GeO_2
sp. w. oxide	4.7	4.703
chloride	$EsCl_4$	$GeCl_4$
boil pnt. chloride	<100°	86°
density chloride	1.9	1.887
flouride	EsF_4	$GeF_4 \cdot 3H_2O$
not gaseous	—	white solid mass
ethyl compound	$EsAe_4$	$Ge(C_2H_5O)_4$
boil. pnt. ethyl compound	160°	160°
sp. w. ethyl compound	0.96	a little <1

*Eka= prefix being the Sanskrit numeral one

periodic law from atomic weights to atomic numbers was accomplished largely through the work of H. G. J. Moseley in 1913.

EVASION OF FALSIFICATION BY REDEFINITION

A critic might object that the shift from atomic weights to atomic numbers is a violation of methodological falsificationism. Evidence was presented against the original periodic law, and the (eventual) response was to redefine the class subject to a periodic variation of properties. But is not such redefinition a nonempirical strategy? Consider the following dialogue:

A: "All members of the National Rifle Association oppose gun registration."

B: "That's not true. Jones is a card-carrying NRA member and he sponsored a gun registration bill in the state legislature."

A: "Your facts are correct, but Jones is not a *true* NRA member."

A appears to be advancing an empirically testable claim about NRA members. However, when confronted with negative evidence, *A* responds by redefining the relevant class to "true NRA members."

There is an obvious difference between the two cases of redefinition. The shift to "atomic number" is sanctioned by a theory about atomic structure, and there are many additional testable consequences of this theory. The shift to "true NRA members," by contrast, has no theoretical basis apart from its necessary condition—opposition to gun registration. What counts, it seems, is not the fact of redefinition, but whether the redefinition leads to further testing.

RESTRICTION OF THE SCOPE OF APPLICATION OF A LAW OR THEORY

In many cases scientists respond to negative evidence not by rejecting the theory in question but by restricting its scope of application. This was the fate of theories about pendulums, the refraction of light, and the behavior of gases.

Galileo claimed in *Two New Sciences* that the period of a simple pendulum is independent of the amplitude of its swings. He insisted that pendulum swings are isochronous even at angles of displacement of 70 or 80 degrees.[5] Subsequent investigations revealed that Galileo was wrong. There are appreciable deviations from isochronicity at high angles of displacement. Scientists responded by restricting the scope of the isochronicity relation. The relation is affirmed to hold only for small angles of displacement.

Snel and Descartes independently maintained that the passage of light from one medium to another obeys the law: sin i / sin $r = k$, where i is the angle of incidence, r is the angle of refraction, and k is a constant whose value is determined by the refractive properties of the two media.

The discovery of double refraction in Iceland spar crystals (a form of calcium calcite) led scientists to restrict the range of application of Snel's law.[6] The law was affirmed to hold for the ordinary ray in cases of double refraction, but not for the extraordinary ray. Given a choice between declaring false Snel's law and restricting its range of application, scientists chose the latter option.

The fate of Boyle's law was similar. Scientists did not reject the law when it was discovered to yield values at variance with empirical data at high temperatures and pressures. Rather, it was acknowledged that the law holds only for gases at moderate temperatures and pressures. In response to experimental data on the behavior of very light bodies (e.g., electrons)

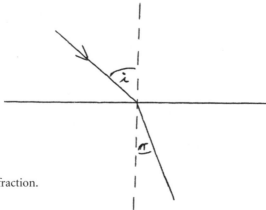

Figure 11. Snel's Law of Refraction.

at very high velocities, most scientists will concede that, strictly speaking, Newtonian mechanics is false. But they do not reject Newtonian mechanics on this account. Rather, they continue to use it after restricting its scope of application to the moderately heavy, relatively slow-moving bodies of everyday experience.

Statistical regularities present further problems for a falsificationist methodology. If new data do not conform to an established statistical regularity, the methodologist may elect not to reject the regularity but to re-calculate the percentage cited in the regularity. For example, suppose a study of men who smoke three packs of cigarettes a day beginning at age twenty shows a mortality rate prior to age fifty of 43%. If a second similar study shows a mortality rate of 31%, the methodologist may either reject the result of one of the two studies or revise the initial percentage in the light of the subsequent study. Statistical hypotheses may be modified repeatedly by recalculation without exposure to falsification at any point.

MODIFICATION OF HYPOTHESES IN THE FACE OF NEGATIVE EVIDENCE

Respect for negative evidence is a requirement of empirical method. But how can one determine whether particular methodological practice manifests this respect? In most cases it is a grey area. An illustration that anchors the black end of the scale is the "Parable of the Gardener" discussed by John Wisdom and Antony Flew.[7] In this parable, Believer and Sceptic assume that the gardener hypothesis is an existentially quantified assertion of the form $(\exists x) [(Gx \bullet Txp) \bullet (y) (Typ \supset y = x)]$ in which $Gx =$ x is a gardener, $Txy =$ x tends y, and $p =$ the plot in question. Believer and Sceptic agree that the hypothesis is that there is a single Gardener involved and that he both has tended the plot and will continue to do so.

Believer and Sceptic come upon a jungle clearing where flowers are growing among the weeds. The Believer announces that "A Gardener tends this plot." The Sceptic proposes a watch. No Gardener is seen. The Sceptic takes this to constitute refutation. But the Believer modifies his original hypothesis to the claim that "An invisible, but otherwise normal, Gardener tends this plot." The Sceptic suggests a patrol by bloodhounds. The hounds remain calm. The Believer responds with a further qualification—"An invisible and odorless, but otherwise normal, Gardener tends this plot." The Sceptic suggests that the area be electrified. No discharge is recorded. The Believer accommodates this evidence by claiming that "It is an invisible, odorless and intangible, but otherwise normal, Gardener who tends this plot." At which point the Sceptic demands to know how this "invisible, odorless and intangible Gardener" differs from no gardener at all.

The Believer's methodology is revealed in his response to negative evidence. His initial hypothesis is that "A Gardener tends this plot." He agrees to a test of this hypothesis. A watch is set. Presumably, he would take an encounter with a being wielding a hoe to support his hypothesis. The Sceptic expects that the Believer would abandon his hypothesis if the watch reveals no activity.

The Believer, however, chooses to respond to negative evidence by modifying the initial hypothesis. He now asserts that "An *invisible*, but otherwise normal, Gardener tends this plot." This move accounts for the negative evidence. Moreover, the Believer agrees to expose the modified hypothesis to a test. A bloodhound patrol is set. The Sceptic expects that the Believer will abandon the modified hypothesis if the bloodhounds remain quiet.

The Believer, however, responds to negative evidence about the bloodhounds' behavior by further modifying the gardener hypothesis. He now asserts that "An invisible, *odorless*, but otherwise normal Gardener tends this plot." The two explorers agree to test this hypothesis by enclosing the area with an electrified fence. The Sceptic expects that the Believer will abandon the "invisible and odorless gardener hypothesis" if no electric discharge is recorded. The Believer, however, responds to negative evidence from the electric meters by making yet another modification of the gardener hypothesis. He now asserts that "An invisible, odorless, *intangible*, but otherwise normal Gardener tends this plot." At this point in the parable the Sceptic concludes that the Believer has abandoned all pretense to be following an empirical methodology. As Flew put it, "A fine brash hypothesis may thus be killed by inches, the death by a thousand qualifications."[8]

But the Believer may argue that his thrice-qualified hypothesis still has life. The Sceptic will require a further test. Suppose the Believer proposes as a test the continued flourishing of the flowers for one month. If the Believer has evidence that, without intervention, weeds soon choke off flowering plants under conditions similar to those in the clearing, then he might claim that a return to the clearing one month hence is a suitable test. And since methodological falsificationism requires only a *pattern* of continued testing, the Believer might claim that he has practiced proper empirical method at all stages of the inquiry.

Flew is correct to warn of a "death by a thousand qualifications." Most observers, myself included, would relegate the Believer's suggestions to the realm of fairy tales at some point in his defense of the gardener hypothesis. But the transition from empirical to nonempirical methodology is not sharp.

One interesting feature of the gardener parable is that at no point does the modification of a hypothesis provide additional test possibilities. Whatever tests are pertinent to hypothesis #2 also are pertinent to hypothesis #1; whatever tests are pertinent to hypothesis #3 also are pertinent to hypothesis #2, and so forth. For example, if baying bloodhounds support the invisible gardener hypothesis, they also support the initial gardener hypothesis. On the contrary, the succession of hypotheses offered by the Believer progressively decreases the range of possible tests. Perhaps it is a mark of acceptable modifications of hypotheses that the range of relevant tests not be decreased thereby.

Karl Popper sought to enforce a restriction of this type. He declared that "as regards auxiliary hypotheses we decide to lay down the rule that only those are acceptable whose introduction does not diminish the degree of falsifiability or testability of the system in question, but, on the contrary, increases it."[9] Popper's criterion is useful only if the comparative "degree of falsifiability" of theories can be estimated. There are some historical episodes for which this is plausible.

The criterion appears to legitimize the transition from atomic-weight periodicity to atomic-number periodicity. The shift of focus to atomic number was based on acceptance of a new theory about the internal structure of atoms. This theory of atomic structure was subjected to numerous new tests. Among these tests were nuclear scattering experiments and the results of X-ray spectroscopy. The range of potential falsifiers is greater for atomic-number periodicity than for atomic-weight periodicity.

By contrast, Popper's criterion disqualifies J. B. Dumas' modification of Prout's hypothesis. Prout had suggested in 1816 that the weights of the various chemical elements are integral multiples of the atomic weight of hydrogen (1.0 gm/gm atom). This hypothesis received support from studies of carbon (12), nitrogen (14), oxygen (16), sulfur (32), and other elements. Unfortunately, the atomic weights of some elements—notably chlorine (35.5)—appeared to be nonintegral. Dumas noted that the atomic weight of chlorine can be accommodated by taking the basic chemical building block to have a weight of 0.5 gm/gm atom. He suggested that Prout's hypothesis be modified so that the weights of chemical elements are multiples of one-half the atomic weight of hydrogen.[10] Dumas' modification reduced the range of potential falsifiers. Atomic-weight values near X.5, where X is an integer, count against Prout's hypothesis, but no longer count against Dumas' modified Proutian hypothesis. Dumas went so far as to suggest that it may be appropriate to take the basic chemical building block to have a weight that is one-fourth the weight of the hydrogen atom. Such a move would further decrease the range of potential falsifiers, since atomic-weight values near X.25, X.50, and X.75, which count against Prout's original hypothesis, do not count against this modified hypothesis.

The above applications of Popper's criterion depend on the specific details of the transitions. Can a general measure of "degrees of falsifiability" be developed? In *The Logic of Scientific Discovery* and other writings Popper sought to achieve this. Unfortunately, this project did not succeed. In his postwar writings Popper emphasized that it is the severity of tests that is important, and he conceded that no universally applicable scale of test severity is available.[11]

POPPER ON SEVERE TESTS

The theoretician will for several reasons be interested in non-refuted theories, especially because some of them *may* be true. He will prefer a non-refuted theory to a refuted one, provided it explains the successes and failures of the refuted theory.

But the new theory may, like all non-refuted theories, be false. The theoretician will therefore try his best to detect any false theory among the set of non-refuted competitors; he will try to "catch" it. That is, he will, with respect to any given non-refuted theory, try to think of cases or situations in which it is likely to fail, if it is false. Thus he will try to construct *severe* tests, and *crucial* test situations. This will amount to the construction of a falsifying law; that is, a law which may perhaps be of such a low level of universality that it may not be able to explain the successes of the theory to be tested, but which will, nevertheless, suggest a *crucial experiment* which may refute, depending on its outcome, either the theory to be tested or the falsifying theory.

By this method of elimination, we may hit upon a true theory. But in no case can the method *establish* its truth, even if it is true; for the number of *possibly* true theories remains infinite, at any time and after any number of crucial tests. (This is another way of stating Hume's negative result.) The actually proposed theories will, of course, be finite in number; and it may well happen that we refute all of them, and cannot think of a new one.

On the other hand, *among the theories actually proposed* there may be more than one which is not refuted at a time *t*, so that we may not know which of these we ought to prefer. But if at a time *t* a plurality of theories continues to compete in this way, the theoretician will try to discover how crucial experiments can be designed between them; that is, experiments which could falsify and thus eliminate some of the competing theories.

The procedure described may lead to a set of theories which are "competing" in the sense that they offer solutions to at least some common problems, although each offers in addition solutions to some problems which it does not share with the others. For although we demand of a new theory that it solves those problems which its predecessor solved and those which it failed to solve, it may of course always happen that two or more new competing theories are proposed such that each of them satisfies these demands and in addition solves some problems which the others do not solve.

At any time *t*, the theoretician will be especially interested in finding the best testable of the competing theories in order to submit it to new tests.

Tests of theories resemble the destruction tests that engineers perform to certify the strengths of construction materials. As such, tests are to be distinguished from mere instances. The discovery of yet another temperate-region raven that is black, for instance, is probably not a "test" of the hypothesis that "all ravens are black." That this new instance of ravenhood is black is, comparatively speaking, unimportant. But suppose several generations of ravens were observed living in polar regions where cosmic-ray intensities—and presumably, mutation frequencies—are higher. Discovery that one of those ravens is black might be regarded as important and hence a test of the hypothesis.

It is difficult to say just what makes a test "severe." One important factor is the extent to which the most general theories are affected. If a negative test result would require a wholesale revision of the most basic theories accepted at the time, then the test qualifies as severe. For instance, the discovery that light from the stars is bent by the sun, as predicted by Einstein's theory of general relativity, is a severe test because a negative test result would have had extensive theoretical repercussions. By contrast, the discovery of an albino raven could be more easily accommodated. The more dramatic the anticipated refutation that is successfully avoided (without adjusting the theory in response to the test result), the more acceptable the theory. The severity of a test depends as well on the accuracy and precision of the experimental data and the relationship of the data to competing theories.

In the absence of a satisfactory quantitative measure of test severity, the best approach is perhaps to identify historical cases for which there is widespread agreement that a severe test has been passed. Persuasive examples of severe tests passed are easy to find. Given the observed motion of the moon, Newton's theory of gravitational attraction requires that the earth be flattened at the poles. The French Academy expedition to Lapland in 1736 established that the earth indeed is flattened at the poles. Newton's theory received credit for having passed an important test. Mendeleyev's theory of the periodic properties of the chemical elements required that there exist an element below silicon in his periodic table, an element of atomic weight near 72, density 5.5, which forms an oxide XO_2 of density 4.7, a sulfide XS_2 insoluble in water but soluble in ammonium sulfide, and a chloride XCl_4 which is a liquid at room temperature with a density of 1.9. The element germanium was discovered in 1886, fifteen years after Mendeleyev had made these predictions. Each prediction was confirmed experimentally. This confirmation clearly should count as an important test passed. Pauli's theory of beta-decay (1930), in which an electron is emitted from a radioactive nucleus, required that there exist a hitherto undiscovered particle. In 1953, direct experimental evidence of the existence of the neutrino was achieved, thereby providing the theory with important support.

It is important that the test result be described without reference to the postulates of the theory under test. This condition is fulfilled for the examples above. The radius of the earth can be described without reference to the Newtonian theory of gravitational attraction. The physical and chemical properties of germanium can be described without reference to the theory that such properties vary periodically with atomic weight. And the types of events that qualify as "passage of neutrinos" can be described without reference to the theory of beta-decay.

Methodological falsificationism forbids a policy of consistent evasion of potentially falsifying evidence. But the evasion of negative evidence at a particular stage of development of a theory may be consistent with a commitment to long-term falsificationism. If one looks to the history of science for guidance, it is clear that, in some cases modifying a theory in the face of falsifying evidence has contributed to progress, whereas in other cases such modification has thwarted progress.

FRUITFUL AND UNFRUITFUL MODIFICATIONS OF THEORIES IN THE HISTORY OF SCIENCE

The following modifications proved fruitful:

1. Leverrier and Adams, confronted with evidence that the orbit of Uranus does not conform to the requirements of Newtonian gravitational theory, posited the existence of a trans-Uranic planet. Johann Galle subsequently identified (1846) a planet—Neptune—at the position on the celestial sphere predicted by Leverrier and Adams.

2. Pauli and Fermi (1934), confronted with evidence that the reaction products in beta-decay possess less energy than the original nucleus, hypothesized that a new particle—the neutrino—carries off just enough energy to ensure energy conservation. The hypothesis was confirmed by the discovery of the neutrino in 1953.

3. Goudsmit and Uhlenbeck (1925), confronted with evidence that the spectral lines of alkali metals—lithium, sodium, potassium, et al.—show a doublet structure, suggested that orbital electrons exist in one of two op-

posed spin states. The hypothesis of electron spin not only accounted for alkali-metal spectra, but also has received support from experimental studies of subatomic particle interactions.

The following modifications, in contrast, did not prove fruitful:

4. Leverrier (1859), confronted with evidence that the orbit of Mercury does not conform to the requirements of Newtonian gravitational theory, posited the existence of a planet (Vulcan) whose orbit is interior to that of Mercury. Extensive observations of the relevant region inside the orbit of Mercury failed to reveal a planet.

5. Clairaut (1748), confronted with evidence that the motion of the moon's line of apsides (the line through its point of closest approach to the earth and its point of greatest distance from the earth) did not agree with predictions derived from Newton's law of gravitational attraction, suggested that the law be replaced by a law of the following form:

$$F = \frac{k_1 m_1 m_2}{r^2} + \frac{k_2 m_1 m_2}{r^4}$$

This suggestion received little support at the time it was made. Clairaut himself subsequently discovered that the discrepancy was due, not to the inadequacy of Newton's law, but to errors of calculation in applying the law. Certain very small second-order products had been eliminated in the original calculations. The cumulative effect of this elimination was to introduce a significant gap between the calculated and the observed motion of the moon.

6. H. A. Lorentz and G. F. Fitzgerald (1889 ff.) attributed the null result of the Michelson-Morley experiment, designed to measure the earth's motion through the aether, to a contraction in length of the instruments used to measure the velocity of light. Scientists gradually lost interest in this proposal because no further applications of the contraction hypothesis were developed. The contraction hypothesis increasingly was perceived to be an *ad hoc* invention to save the aether theory from falsification.

Judgments about "fruitfulness" require historical inquiry. It is only from the standpoint of subsequent developments that evasive actions can be judged to be contributions to scientific progress.

3. THE REPLACEMENT
   OF THEORIES

7. Whewell on Scientific Progress

The preceding survey has shown that confrontations between the implications of theory and observational evidence are not always resolved in favor of the evidence. Various strategies have been developed to avoid falsification. On some occasions pursuit of these strategies has contributed to the development of science. On other occasions attempts to protect a theory proved counterproductive.

Since theories do replace other theories within the history of science, the question arises whether there is some set of conditions present whenever T_2 is determined to be superior to T_1. An examination of the history of science might reveal a pattern displayed by those theory replacements that turned out to be fruitful.

William Whewell undertook such an examination in the mid-nineteenth century. The pattern Whewell professed to find throughout the history of science is the systematic incorporation of prior results into present theories. Whewell became the leading spokesman for a "growth by incorporation" view which emphasizes continuity in science rather than revolutionary confrontation and overthrow.

The history of science that Whewell surveyed is itself an interpretation of the available records. In order to write a history of science one must judge which human activities count as the practice of science and what types of factors are important in the development of science. The first problem is a problem of demarcation. Decisions must be made about the inclusion or exclusion of astrology, alchemy, natural history, and philosophical theories about the nature of the universe. The second problem concerns the relative importance of diverse aspects of the scientific enterprise. The historian of science must assess the relative importance of imaginative speculation, the accumulation of observational evidence, the stimulus of political or economic conditions, and the force of personality.

Whewell was well aware that it is necessary to adopt a methodological orientation in order to write a history of science. His own *History of the Inductive Sciences* (first edition 1837)[1] is based on two important assumptions about the nature of science. The first assumption is that science comprises a set of individual subdisciplines—astronomy, optics, botany, crystallography, and so on—each with a distinctive subject matter, basic predicates, and fundamental axioms. The second assumption is that scientific knowledge arises from an interaction of "facts" and "ideas."

Roughly speaking, "facts" are reports about what is observed, and "ideas" are principles that govern relations among facts. Whewell emphasized that there are no "pure facts" independent of all ideas. Any description of an object or process involves ideas of space, time, or number. Consequently, even the simplest facts have a theoretical component. Whewell included among ideas both concepts applicable to facts in every science— for example, space, time, and cause—and concepts applicable only to facts in a particular science, e.g., polarization (optics), vital forces (biology), and elective affinity (chemistry).

Whewell maintained that theory formation is a superposition of an idea, or ideas, upon facts, in such a way that the facts are "seen in a new light."[2] Thus Kepler superimposed the idea of an elliptical orbit upon facts about the positions of a planet against the zodiac. Kepler redescribed the data as "points on elliptical orbits," thereby creating new candidates for factual status. Subsequently, Newton superimposed the ideas of rectilinear inertial motion and gravitational attraction upon the "facts" of the Keplerian orbits. Whewell's distinction between fact and theory is thus a relative distinction. From Newton's point of view, Kepler's elliptical orbits are facts to be explained; from the point of view of Kepler's contemporaries, his elliptical orbits are theoretical constructions superimposed upon the basic facts of planetary positions.

Given these basic assumptions, there are three ways in which science may develop over time. (1) There may occur a "decomposition of facts" that reveals fundamental spatiotemporal dependencies within the facts. Such decomposition may contribute to the growth of science by rendering the facts susceptible to incorporation into theories. (2) There may occur an "explication of conceptions" in which vaguely apprehended notions become clarified by means of the introduction of appropriate definitions.[3] And (3) there may occur a "colligation of facts" in which the facts are bound together by reference to some theoretical relationship among ideas.

Whewell surveyed the history of science he had created upon application of the fact-idea polarity and concluded that it conforms to a pattern. Theories within a particular science flow into their successors as streams flow into rivers. A "tributary-river" pattern is the characteristic feature of the growth of a science (figure 12).[4]

In the case of the history of astronomy, for instance, Whewell identified a number of "tributaries" that became incorporated into the "river" of Newton's theory of gravitational attraction. Among these tributaries are the facts that

1. The planets obey (roughly) Kepler's three laws;

2. The satellites of Jupiter and Saturn also obey Kepler's laws;

3. The motions of Jupiter and Saturn display inequalities that vary with their relative positions with respect to one another and the sun;

4. The moon moves in an elliptical orbit with a variable axis and eccentricity;

5. The tides vary with the relative positions of the moon and the sun;

6. The weights of bodies decrease as they are transported from polar regions toward the equator; and

7. There is a slow precession of the equinoxes.

Each tributary, in turn, is a product of the confluence of other streams.

Copernicus is often said to have inaugurated a revolution in astronomy, a revolution to which Kepler and Galileo made important contributions. Whewell preferred to emphasize continuity, pointing out that geocentrists such as Ptolemy and Brahe created the factual base upon which heliostatic theories were formulated. He claimed that criteria of theory replacement ought to reflect the continuity displayed in the history of science.

Whewell labeled the appropriate criterion of theory replacement the "consilience of inductions." Consilience requires, as necessary conditions

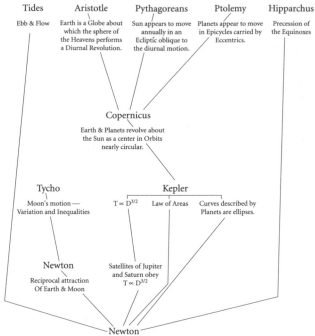

Figure 12. Whewell's Tributary-River Pattern of the History of Astronomy (in part). William Whewell, *Philosophy of the Inductive Sciences,* 2nd ed., 1847 (London: Cass, 1967), vol. 2, p. 118.

of the acceptability of the replacement of one theory by another, that (1) the successor theory be consistent; (2) the successor theory be more inclusive than its predecessor (the range of facts subsumed by the successor theory must be greater than the range subsumed by its predecessor); and (3) the increase in subsumptive power be accompanied by a gain in simplicity.

Whewell's insistence that consilience requires that ideas be superimposed upon facts in such a way that the facts are interpreted in a new way. The mere conjunction of previously successful theories does not qualify as consilience. To satisfy condition 3 above, a successor theory must integrate, and not just conjoin, the relevant facts.

It is one thing to require that acceptable theories be "simple" and quite another thing to determine what counts as "simplicity." Unfortunately Whewell did not specify a procedure for deciding when a successor theory qualifies as "simple." To give content to this necessary condition of consilience, it is necessary to examine Whewell's specific judgments about the history of science.

Whewell praised Newton's theory of gravitational attraction as the supreme example of the achievement of consilience within the history of science. Newton subsumed facts about planetary motions, the tides, the motion of pendulums, and so forth, by means of a theory about central forces, acceleration, and inertial motion. Universal gravitational attraction is a simple, but immensely powerful, idea. It explains not only why lower-level correlations hold, but also why they hold only approximately. Deviations from Kepler's laws by Jupiter, Saturn, and the moon are accounted for by applications of the theory of gravitational attraction itself. This is accomplished by taking account of the relevant conditions under which the theory is applied: the influence of a third body in the case of the motions of Jupiter and Saturn, and the asymmetry of the earth in the case of the motion of the moon. It would seem from this analysis that a theory qualifies as simple when its explanatory power is increased without introducing new assumptions into the theory itself.

But scientific theories often are modified to accommodate new facts. Whewell acknowledged this. He was prepared to issue judgments of relative simplicity on the modifications of competing theories. He observed that Newton's corpuscular theory of light and the wave theory of Young and Fresnel both had been modified to fit the facts of double refraction, diffraction, and polarization.

According to Whewell, supporters of the wave theory succeeded in accounting for double refraction, diffraction, and polarization by the "simple and harmonious" hypothesis that the propagation of light is a transmission of transverse waves. Supporters of the corpuscular theory, by contrast, added a hypothesis about different forces along different axes of a crystal to account for double refraction; a hypothesis about the "asymmetrical sides" of a corpuscle to account for polarization; and a hypothesis about "fits of easy transmission and reflection" (Newton) to account for the colors produced by the contact of thin plates. In Whewell's judgment, these modifications of the corpuscular theory were complex and disharmonious. The wave theory displayed consilience over time. The corpuscular theory did not.

Unfortunately, Whewell did not formulate a general criterion of comparative simplicity. Consequently, he provided no means to determine, for an arbitrarily selected case of theory modification, whether conditions necessary for consilience are present. However, Whewell did specify a sufficient condition of consilience. Consilience is achieved whenever a theory is modified so as to reveal "undesigned scope." He declared that "the evidence in favour of our induction is of a much higher and more forcible character when it enables us to explain and determine cases of a kind different from those which were contemplated in the formation of our hypothesis."[5] Whewell's contemporary John Herschel placed a similar emphasis on undesigned scope.[6]

Whewell defended the value of a display of undesigned scope by an appeal to the history of science. He claimed that every theory that unexpectedly accounts for a new range of phenomena, without specific adjustment to do so, has passed the test of survival. He insisted that if a theory "of itself and without adjustment for the purpose, gives us the rule and reason of a class of facts not contemplated in its construction, we have a criterion of its reality, which has never yet been produced in favour of a falsehood."[7] Whewell thus appealed to the history of science to warrant the achievement of undesigned scope as a sufficient condition of scientific progress. He invited those who disagree to specify a case in which undesigned scope is realized, but the theory in question later is shown to be false.

Herschel and Whewell held that a theory displays undesigned scope provided that it has an application that is (1) unexpected, and (2) different in kind from those taken into account when the theory was formulated. Unfortunately, neither Herschel nor Whewell set forth an objective standard to determine when cases are "different in kind."

One case discussed by both Herschel and Whewell is Laplace's extension of his theory of heat transfer to the propagation of sound.[8] Laplace succeeded in explaining a hitherto puzzling discrepancy between the calculated and measured values of the velocity of sound. Prior calculations had omitted to take account of the heat generated upon the compression of the air caused by the passage of a sound wave. Herschel and Whewell maintained that, since Laplace formulated his theory of heat transfer without consideration of the properties of sound, this application of the theory was a display of undesigned scope.

However, one could argue that this was not an application to a case "different in kind." Of course the compression of an elastic medium generates heat. The theory of heat transfer applied to sound all along. Laplace was just the first to recognize this consequence of his theory. On this interpretation, Laplace's result may have been "unexpected," but it was not an extension to a "new kind" of phenomena.

In the absence of an objective standard to determine whether a theory application is "different in kind," one must assess undesigned scope by reference to psychological impact. If many scientists find a successful theory application unexpected, then this may be taken to establish undesigned scope. This "unexpectedness" standard is vague. Moreover, it is difficult to reconstruct scientists' reactions to theory applications from the available records. But given a reasonably strict standard of "unexpectedness", the achievement of undesigned scope is rare within the history of science. Thus, even if undesigned scope (measured by unexpectedness) is a sufficient condition of consilience, very few theories qualify as consilient on this ground. Most questions about progressive theory change must be answered by reference to the vague requirements of inclusiveness and simplicity.

There is a certain circularity in Whewell's analysis of scientific progress. Whewell began the analysis by positing a fact-idea distinction and the Aristotelian conception of a series of distinct sciences, each with a set of axioms that state relations among the "fundamental ideas" of that science. He then noted that the history of science reveals that scientific progress is a continuing superposition of increasingly precise ideas upon an expanding factual base. Past results are subsumed and reinterpreted by present theories. Whewell concluded that the "consilience of inductions" is the evaluative criterion revealed by the history of science.

Whewell's analysis is not viciously circular. The initial methodological assumptions do not imply the consilience criterion. Nevertheless, to make these initial methodological assumptions is to stack the case in favor of an emphasis on inclusiveness and simplicity.

8. Kuhn on Theory Replacement

A THEORY IS REJECTED ONLY WHEN A
COMPETING THEORY IS AVAILABLE

Thomas Kuhn complained that Whewell had overemphasized growth by incorporation. Kuhn insisted that the development of science is characterized not just by accretion but also by revolutionary confrontations in which one theory is vanquished by another. A revolutionary confrontation takes place in response to the recognition that a previously successful theory has failed to explain phenomena that scientists believe it ought to explain. The accumulation of anomalies of this type provides a stimulus for the invention of competing theories.

The physicist Paul Dirac, looking back on developments that led to the formulation of quantum mechanics, focused on the anomalies that beset classical mechanics.[1]

DIRAC ON ANOMALIES UNEXPLAINED BY
CLASSICAL MECHANICS

The necessity for a departure from classical mechanics is clearly shown by experimental results. In the first place the forces known in classical electrodynamics are inadequate for the explanation of the remarkable stability of atoms and molecules, which is necessary in order that materials may have any definite physical and chemical properties at all. The introduction of new hypothetical forces will not save the situation, since there exist general principles of classical mechanics, holding for all kinds of forces, leading to results in direct disagreement with observation. For example, if an atomic system has its equilibrium disturbed in any way and is then left alone, it will be set in oscillation and the oscillations will get impressed on the surrounding electromagnetic field, so that their frequencies may be observed with a spectroscope. Now whatever the laws of force governing the equilibrium, one would expect to be able to include the various frequencies in a scheme comprising certain fundamental frequencies and their harmonics. This is not observed to be the case. Instead, there is observed a new and unexpected connexion between the frequencies, called Ritz's Combination Law of Spectroscopy, according to which all the frequencies can be expressed as differences between certain terms, the number of terms being much less than the number of frequencies. This law is quite unintelligible from the classical standpoint.

One might try to get over the difficulty without departing from classical mechanics by assuming each of the spectroscopically observed frequencies to be a fundamental frequency with its own degree of freedom, the laws of force being such that the harmonic vibrations do not occur. Such a theory will not do, however, even apart from the fact that it would give no explanation of the Combination Law, since it would immediately bring one into conflict with the experimental evidence on specific heats. Classical statistical mechanics enables one to establish a general connexion between the total number of degrees of freedom of an assembly of vibrating systems and its specific heat. If one assumes all the spectroscopic frequencies of an atom to correspond to different degrees of freedom, one would get a specific heat for any kind of matter very much greater than the observed value. In fact the observed specific heats at ordinary temperatures are given fairly well by a theory that takes into account merely the motion of each atom as a whole and assigns no internal motion to it at all.

This leads us to a new clash between classical mechanics and the results of experiment. There must certainly be some internal motion in an atom to account for its spectrum, but the internal degrees of freedom, for some classically inexplicable reason, do not contribute to the specific heat. A similar clash is found in connexion with the energy of oscillation of the electromagnetic field in a vacuum. Classical mechanics requires the specific heat corresponding to this energy to be infinite, but it is observed to be quite finite. A general conclusion from experimental results is that oscillations of high frequency do not contribute their classical quota to the specific heat.

As another illustration of the failure of classical mechanics we may consider the behaviour of light. We have, on the one hand, the phenomena of interference and diffraction, which can be explained only on the basis of a wave theory; on the other, phenomena such as photo-electric emission and scattering by free electrons, which show that light is composed of small particles. These particles, which are called photons, have each a definite energy and momentum, depending on the frequency of the light, and appear to have just as real an existence as electrons, or any other particles known in physics. A fraction of a photon is never observed.

Experiments have shown that this anomalous behaviour is not peculiar to light, but is quite general. All material particles have wave properties, which can be exhibited under suitable conditions. We have here a very striking and general example of the breakdown of classical mechanics, not merely an inaccuracy in its laws of motion, but an inadequacy of its concepts to supply us with a description of atomic events.

The necessity to depart from classical ideas when one wishes to account for the ultimate structure of matter may be seen, not only from experimentally established facts, but also from general philosophical grounds. In a classical explanation of the constitution of matter, one would assume it to be made up of a large number of small constituent parts and one would postulate laws for the behaviour of these parts, from which the laws of the matter in bulk could be deduced. This would not complete the explanation, however, since the question of the structure and stability of the constituent parts is left untouched. To go into this question, it becomes necessary to postulate that each constituent part is itself made up of smaller parts, in terms of which its behaviour is to be explained. There is clearly no end to this procedure, so that one can never arrive at the ultimate structure of matter on these lines. So long as *big* and *small* are merely relative concepts, it is no help to explain the big in terms of the small. It is therefore necessary to modify classical ideas in such a way as to give an absolute meaning to size.

At this stage it becomes important to remember that science is concerned only with observable things and that we can observe an object only by letting it interact with some outside influence. An act of observation is thus necessarily accompanied by some disturbance of the object observed. We may define an object to be big when the disturbance accompanying our observation of it may be neglected, and small when the disturbance cannot be neglected. This definition is in close agreement with the common meanings of big and small.

It is usually assumed that, by being careful, we may cut down the disturbance accompanying our observation to any desired extent. The concepts of big and small are then purely relative and refer to the gentleness of our means of observation as well as to the object being described. In order to give an absolute meaning to size, such as is required for any theory of the ultimate structure of matter, we have to assume that there is a limit to the fineness of our powers of observation and the smallness of the accompanying disturbance—a limit which is inherent in the nature of things and can never be surpassed by improved technique or increased skill on the part of the observer. If the object under observation is such that the unavoidable limiting disturbance is negligible, then the object is big in the absolute sense and we may apply classical mechanics to it. If, on the other hand, the limiting disturbance is not negligible, then the object is small in the absolute sense and we require a new theory for dealing with it.

A consequence of the preceding discussion is that we must revise our ideas of causality. Causality applies only to a system which is left undisturbed. If a system is small, we cannot observe it without producing a serious disturbance and hence we cannot expect to find any causal connexion between the results of our observations. Causality will still be assumed to apply to undisturbed systems and the equations which will be set up to describe an undisturbed system will be differential equations expressing a causal connexion between conditions at one time and conditions at a later time. These equations will be in close correspondence with the equations of classical mechanics, but they will be connected only indirectly with the results of observations. There is an unavoidable indeterminacy in the calculation of observational results, the theory enabling us to calculate in general only the probability of our obtaining a particular result when we make an observation.

Scientists did not abandon classical mechanics in the face of the anomalies listed by Dirac until alternative theories were formulated. Kuhn claimed that the unwillingness to reject a theory until an alternative is available is a general feature of the history of science. He claimed that, in the case of high-level theories that achieve status as "paradigms," "no process yet discovered by the historical study of scientific development at all resembles the methodological stereotype of falsification by direct comparison with nature."[2] According to Kuhn, the replacement of high-level theory T_1 always involves the triadic relationship $[T_1, e, T_2]$, where T_2 is a competing theory. He declared that "once it has achieved the status of paradigm, a scientific theory is declared invalid only if an alternative candidate is available to take its place."[3]

Imre Lakatos echoed Kuhn's claim. According to Lakatos, "contrary to naive falsificationism, no experiment, experimental report, observation statement or well-corroborated low-level falsifying hypothesis alone can lead to falsification. There is no falsification before the emergence of a better theory."[4]

Certain episodes from the history of science provide support for the Kuhn-Lakatos thesis on the abandonment of theories. For instance, late nineteenth-century scientists did not abandon Newtonian gravitational theory in the face of repeated failures to account for the anomalous precession of Mercury's orbit. It was not until the formulation of Einstein's general theory of relativity (1916) that the Newtonian theory was rejected. (Scientists recognized that the general theory of relativity is the superior theory. Strictly speaking, the Newtonian theory is false. Nevertheless, they continued to use Newtonian theory to describe the motions of moderately large, slowly moving bodies under many sets of conditions.)

Kuhn restricted the theory abandonment thesis to high-level theories. There is no dispute about the rejection of low-level hypotheses. Such hypotheses often are rejected upon receipt of negative evidence, even though no competing hypothesis is available at the time. For example, the hypothesis that this gas is oxygen will be rejected if it fails to support the combustion of a burning match, even though no other hypothesis has been proposed.

But even if we restrict the thesis to high-level theories, or "paradigms," there are exceptions to the generalization that no theory is rejected until a viable competitor is available. A good example is the abandonment of the theory that mass is conserved in all physical interactions. This principle was abandoned in the face of evidence about radioactive decay processes. At the time no competing theory was available. Of course, a new principle soon was proposed—the conservation of mass-energy.

A supporter of the Kuhn-Lakatos thesis might object that the principle of conservation of mass was reinterpreted rather than abandoned. The reinterpretation is based on the presumed interconvertibility of mass and energy. But this interconvertibility is a consequence of the theory of special relativity (1905), and this theory had not been formulated at the time of the discovery of mass loss in radioactive decay processes. The principle of conservation of mass-energy was not available as a competing principle at the time the facts about mass relations in radioactive decay became known. Of course, there always is a competing theory available in a test situation—the negation of the theory tested. But the Kuhn-Lakatos thesis becomes a vacuous truth if $T_2 = \sim T_1$ in the triad $[T_1, e, T_2]$.

The Kuhn-Lakatos thesis does not fit the demise of the principle of parity either. This principle, which stipulates that all physical processes display mirror symmetry, was widely accepted by scientists in the 1940s and early 1950s. The principle was applied successfully in the analysis of atomic spectra. However, extension of the principle from parity in atomic interactions to parity in nuclear interactions led to a problem. In 1956, C. N. Yang and T. D. Lee predicted that parity is not conserved in certain "weak interactions"—decay processes involving mesons. Experiments confirmed this prediction and the principle of parity was abandoned. No well-developed competing theory of weak interactions was available at that time.

C. N. YANG ON SYMMETRY IN NUCLEAR REACTIONS

One of the symmetry principles, the symmetry between the left and the right, has been discussed since ancient times. The question whether nature exhibits such symmetry was debated at great length by philosophers of the past. In daily life, of course, right and left are quite different from each other. In biological phenomena, it was known since Pasteur's work in 1848 that organic compounds appear oftentimes in the form of only one of two kinds, both of which, however, occur in inorganic processes and are mirror images of each other. In fact, Pasteur had considered for a time the idea that the ability to produce only one of the two forms was the very prerogative of life.

The laws of physics, however, have always shown complete symmetry between the left and the right. This symmetry can also be formulated in quantum mechanics as a conservation law called the conservation of parity, which is completely identical to the principle of right-left symmetry. The first formulation of the concept of parity was due to E. P. Wigner. It rapidly became very useful in the analysis of atomic spectra. The concept was later extended to cover phenomena in nuclear physics and the physics of mesons and strange particles. One became accustomed to the idea of nuclear parities as well as atomic parities, and one discussed and measured the parities of mesons. Throughout these developments the concept of parity and the law of parity conservation proved to be extremely fruitful, and the success had in turn been taken as a support for the validity of right-left symmetry in physical laws.

In the years 1954–1956 a puzzle called the θ-τ puzzle developed. The θ and τ mesons are today known to be the same particle, usually called K. In those years, however, one only knew that there were particles that disintegrate into two π mesons and particles that disintegrate into three π mesons. They were called respectively θ's and τ's, the τ being the name given to it by Powell in 1949. As time went on, measurements became more accurate and the increasing accuracy brought out more and more clearly a puzzlement. On the one hand it was clear that θ and τ had very accurately the same mass. They were also found to behave identically in other respects. So it looked as if θ and τ were really the same particle disintegrating in two different ways. On the other hand, increasingly accurate experiments also showed that θ and τ did not have the same parity and could not therefore be the same particle.

The resolution of the puzzle lay in a change in the concept of right-left symmetry. In the summer of 1956, T. D. Lee and I examined the then existing experimental foundation of this concept and came to the conclusion that, contrary to generally held belief, no experimental evidence of right-left symmetry actually existed for the weak interactions. If right-left symmetry does not hold for the weak interactions, the concept of parity is inapplicable to the decay mechanism of the θ and τ particles and they could therefore be one and the same particle, as we now know they are.

As a possible way out of the θ-τ puzzle, it was suggested that one should test experimentally whether right-left symmetry is violated for the weak interactions. The principle of the test is very simple: two sets of experimental arrangements which are mirror images of each other are set up. They must contain weak interactions and they must not be identical to each other. One then examines whether the two arrangements always give the same results. If they do not, one would have an unambiguous proof of the violation of right-left symmetry in this experiment. In Figure 34 [figure 13] the first such experiment, performed by C. S. Wu, E. Ambler, R. W. Hayward, D. D. Hoppes, and R. P. Hudson in 1956, is schematically illustrated. The cobalt nuclei disintegrate by weak interactions and the disintegration products are counted. Notice that the currents flowing in the loops are very essential elements of the experiment. Without these currents the two arrangements on the two sides of the imagined mirror would have been identical and would have always given the same results. To make the influence of the currents felt by the cobalt nuclei, however, it was necessary to eliminate the

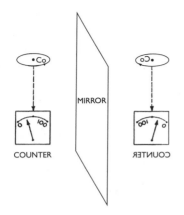

Figure 13. Yang on Weak-Interaction Asymmetry. Chen Ning Yang, *Elementary Particles* (Princeton, N.J.: Princeton University Press, 1962), p. 56. © 1961 Princeton University Press, 1989 renewed PUP. Reprinted by permission of Princeton University Press.

disturbance on the cobalt produced by thermal agitations. The experiments had to be done, therefore, at extremely low temperatures of the order of less than 0.1 degrees absolute.

The result of the experiment was that there was a very large difference in the readings of the two meters shown in Figure 34 [13]. Since behavior of the other parts of the apparatus observes right-left symmetry, the asymmetry must be attributed to the disintegration process of cobalt, which is due to a weak interaction.[5]

The career of the Laplacian program for science is a third episode that fails to fit the Kuhnian thesis. In the period around 1800, Pierre-Simon Laplace sought to implement a program first suggested by Newton. The program was to explain all physical interactions by reference to attractive and repulsive forces that act along a line joining the centers of mass of the bodies affected. Most Continental scientists of the time endorsed the Laplacian program.[6]

The model for the program was Newton's successful applications of a $1/r^2$ central force law to bodies in the solar system. Laplace and his followers sought to uncover the distance-dependent central forces responsible for electrical, magnetic, and chemical effects. They attributed central-force fields to "particles of imponderable matter" (light, heat, and electricity) as well as to particles of ponderable matter. A distant goal of the program was to attribute respiration, digestion, and other life processes to the operation of central forces between particles. Laplace himself contributed important studies of optical refraction and capillary action.

The Laplacian program was an appealing program for scientific inquiry. Unfortunately, it soon was shown to be unrealizable. H. C. Oersted showed in 1820 that certain electromagnetic effects do not conform to the central-force model. Oersted placed a magnetic needle, free to pivot in a horizontal plane, beneath a straight stretch of a current-carrying wire. The needle took a position perpendicular to that of the wire. When the direction of current flow through the wire was reversed, the needle swung to a new position 180 degrees from its initial position. And when the needle was placed above rather than below the wire, its position rotated 180 degrees. Oersted concluded that the magnetic field around the wire is circular at right angles to the flow of particles in the wire. Electromagnetic interaction is not a central-force interaction. Oersted's result was quickly repeated and acknowledged by other investigators.

OERSTED ON ELECTROMAGNETIC EFFECTS

Let the opposite poles of the galvanic apparatus be joined by a metallic wire, which, for brevity, we will call hereafter the joining conductor or else the joining wire. To the effect, however, which takes place in this conductor and surrounding space, we will give the name of electric conflict.

Let the rectilinear part of this wire be placed in a horizontal position over the magnetic needle duly suspended, and parallel to it. If necessary, the joining wire can be so bent that the suitable part of it may obtain the position necessary for the experiment. These things being thus arranged, the magnetic needle will be moved, and indeed, under that part of the joining wire which receives electricity most immediately from the negative end of the galvanic apparatus, will decline towards the west.

If the distance of the joining wire from the magnetic needle does not exceed ¾ of an inch, the declination of the needle makes an angle of about 45°. If the distance is increased the angles decrease as the distances increase. The declination, however, varies according to the efficiency of the apparatus.

The joining wire can change its place either eastward or westward, provided it keeps a position parallel to the needle, without any other change of effect than as respects magnitude; and thus the effect can by no means be attributed to attraction, for the same pole of the magnetic needle which approaches the joining

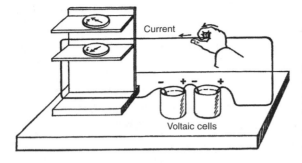

Figure 14. Oersted's Experiment. Guy Omer Jr., Harold L Knowles, Belvey W. Mundy, and Herbert Yoho, *Physical Science: Men and Concepts* (Boston: D. C. Heath, 1962), p. 530.

wire while it is placed at the east side of it ought to recede from the same when it occupies a position at the west side of it if these declinations depended upon attractions or repulsions.

The joining conductor may consist of several metallic wires or bands connected together. The kind of metal does not alter the effects, except, perhaps, as regards quantity. We have employed with equal success wires of platinum, gold, silver, copper, iron, bands of lead and tin, a mass of mercury. A conductor is not wholly without effect when water interrupts, unless the interruption embraces a space of several inches in length.

The effects of the joining wire on the magnetic needle pass through glass, metal, wood, water, resin, earthenware, stones: for if a plate of glass, metal, or wood be interposed, they are by no means destroyed, nor do they disappear if plates of glass, metal, and wood be simultaneously interposed; indeed, they seem to be scarcely lessened. The result is the same if there is interposed a disc of amber, a plate of porphyry, an earthenware vessel, even if filled with water. Our experiments have also shown that the effects already mentioned are not changed if the magnetic needle is shut up in a copper box filled with water. It is unnecessary to state that the passing of the effects through all these materials in electricity and galvanism has never before been observed. The effects, therefore, which take place in electric conflict are as different as possible from the effects of one electric force or another.

If the joining wire is placed in a horizontal plane under the magnetic needle, all the effects are the same as in the plane over the needle, only in an inverse direction, for the pole of the magnetic needle under which is that part of the joining wire which receives electricity most immediately from the negative end of the galvanic will decline towards the east.[7]

Whether or not rejection of the Laplacian program is a countercase to Kuhn's thesis depends on the extent to which the rejection was influenced by available alternative theories. At the time of Oersted's discovery, there were high-level theories in competition with the Laplacian program. Among the non-Laplacian alternatives were Fourier's theory of heat conduction, a mathematical theory that does not invoke central forces between particles, and Fresnel's wave theory of light.[8]

The wave theory of light made significant gains in the mid-nineteenth century at the expense of corpuscular theories consistent with the Laplacian program. However, the principal reason that the Laplacian program was rejected was Oersted's demonstration of the existence of non-central force fields. Again, the Kuhnian thesis about theory rejection does not fit the historical record.

THE LIMITATIONS OF OBJECTIVE STANDARDS OF THEORY APPRAISAL

It is true that Kuhn characterized the experience of high-level theory replacement as a "gestalt-shift" in which phenomena are "seen in a new way." In a much-quoted passage Kuhn wrote, "in a sense that I am unable to explicate further, the proponents of competing paradigms practice their trades in different worlds. One contains constrained bodies that fall slowly, the other pendulums that repeat their motions again and again. In one, solutions are compounds, in the other mixtures. One is embedded in a flat, the other in a curved matrix of space. Practicing in different worlds the two groups of scientists see different things when they look from the same point in the same direction."[9]

Nevertheless, there are objective standards for theory appraisal. First and foremost, a replacement theory must resolve at least some of the anomalies that beset its predecessor. But what counts as "resolving" an anomaly remains to be specified. Standards of resolution may themselves be paradigm dependent. Supporters of an entrenched paradigm may not agree that its anomalies have been resolved by a successor paradigm.

Jack Meiland has suggested a Kuhnian criterion of acceptability for paradigm replacement: "P_{n+1} is superior to P_1 if P_{n+1} solves more of the problems it generates (according to the standard of problem-solution of P_{n+1}) than P_n solves of the problems it generates (according to the standard of problem-solution of P_n)."[10] This is an objective criterion of theory replacement. There are problems, however. The criterion leaves unspecified how problems are to be individuated. Disputes may arise, even within a given paradigm, whether to count p_1 and p_2 as two separate problems or mere aspects of the same problem. Moreover, on Meiland's criterion, one paradigm may be superior to a second simply because its standard of problem solution is the more lenient.

In *The Essential Tension*, Kuhn presented a list of criteria of acceptability which he maintained have been applied, and should be applied, by scientists. These criteria are "accuracy, consistency, scope, simplicity and fruitfulness." Kuhn conceded that there is no general agreement on the weighting of these criteria. He noted as well that "when deployed together, they repeatedly prove to conflict with one another; accuracy may, for example, dictate the choice of one theory, scope the choice of its competitor."[11]

Kuhn emphasized that, because of these difficulties, his list of criteria provide "an insufficient basis for a shared algorithm of choice."[12] He concluded that, since no algorithm of theory choice is available, due attention must be paid to the role played by "idiosyncratic factors dependent on individual biography and personality."[13]

9. Lakatos on Progressive Research Programs

SCIENTIFIC RESEARCH PROGRAMS

Kuhn's work shifted attention from the confrontation between a theory and evidence to the replacement of one theory by another. Imre Lakatos maintained that the replacement of theory T_{n-1} by theory T_n is justified provided that

1. T_n accounts for the previous successes of T_{n-1};
2. T_n has greater empirical content than T_{n-1}; and
3. Some of the excess content of T_n has been corroborated.[1]

One historical episode that fulfills these requirements is the transition from Bohr's theory of the hydrogen atom to Sommerfeld's theory. Sommerfeld assigned two spin orientations to the hydrogen electron. These two orientations are states of different energy.

Sommerfeld's theory accounted for the prior successes of the Bohr

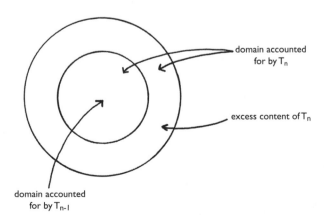

Figure 15. Lakatos's Criterion Incorporation with Corroborated Excess Content.

domain accounted for by T_n

excess content of T_n

domain accounted for by T_{n-1}

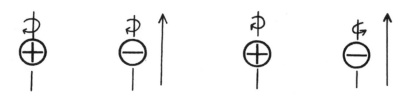

Figure 16. Electron Spins in the Hydrogen Atom.

theory. In addition, it explained, as the Bohr theory did not, the observed fact that the spectral lines of hydrogen occur as closely spaced doublets. Sommerfeld's theory achieved corroborated excess content.

A second historical episode in which a prior theory was incorporated and excess content corroborated is the transition from Mendeleyev's theory of the periodic table to Moseley's theory. Mendeleyev's theory correlates the periodic variation of the chemical properties of the elements with their atomic weights. Mendeleyev's theory accounted for many observed periodicities. But there were anomalies. In order to place iodine in its "proper" group with chlorine and bromine, Mendeleyev reversed the positions by weight of iodine and tellurium. (See chapter 6 above.) This was a violation of his own periodic law to which he appealed in the overall design of the table. A segment of Mendeleyev's periodic table of 1879 follows:[2]

Group VI	Group VII
O=16	F=19
S=32	Cl=35.5
Se=79	Br=80
Te=127.6	I=126

Moseley's theory correlates the periodic variation of the chemical properties of the elements with their atomic numbers (the number of positively charged protons in the nucleus). It accounts for the successes achieved by Mendeleyev's theory. In addition, Moseley's arrangement places iodine in the correct group, because the atomic number of iodine (53) is greater than that of tellurium (52). Moseley's theory thereby incorporates Mendeleyev's theory while corroborating excess content, qualifying it as a warranted successor theory on Lakatos's criterion.

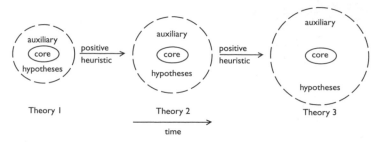

Figure 17. A Lakatosian Scientific Research Program. John Losee, *Philosophy of Science and Historical Enquiry* (Oxford: Clarendon Press, 1987), p. 92.

Lakatos emphasized that the criterion of incorporation with corroborated excess content is to be applied to successive theories within "scientific research programs." According to Lakatos, a scientific research program consists of a central core of axioms and principles and an evolving collection of auxiliary hypotheses adopted in the course of applying the core.[3] The central core is taken to be inviolable by those working within the research program.

LAKATOS ON PROGRESSIVE AND DEGENERATING SCIENTIFIC RESEARCH PROGRAMS

A research programme is said to be progressing as long as its theoretical growth anticipates its empirical growth, that is as long as it keeps predicting novel facts with some success ("progressive problemshift"); it is stagnating if its theoretical growth lags behind its empirical growth, that is as long as it gives only post-hoc explanations either of chance discoveries or of facts anticipated by, and discovered in, a rival programme ("degenerating problemshift"). If a research programme progressively explains more than a rival, it "supersedes" it, and the rival can be eliminated (or, if you wish, "shelved").

(Within a research programme a theory can only be eliminated by a better theory, that is, by one which has excess empirical content over its predecessors, some of which is subsequently confirmed. And for this replacement of one theory by a better one, the first theory does not even have to be "falsified" in Popper's sense of the term. Thus the progress is marked by instances verifying excess content rather than by falsifying instances; empirical "falsification" and actual "rejection" become independent. Before a theory has been modified we can never know in what way it had been "refuted", and some of the most interesting modifications are motivated by the "positive heuristic" of the research programme rather than by anomalies. This difference alone has important consequences and leads to a rational reconstruction of scientific change very different from that of Popper's.)

It is very difficult to decide, especially since one must not demand progress at each single step, when a research programme has degenerated hopelessly or when one of two rival programmes has achieved a decisive advantage over the other. In this methodology, as in Duhem's conventionalism, there can be no instant—let alone mechanical—rationality. Neither the logician's proof of inconsistency nor the experimental scientist's verdict of anomaly can defeat a research programme in one blow. One can be "wise" only after the event.[4]

Lakatos cited post-Newtonian planetary astronomy as a historically important scientific research program. In this Newtonian research program the three axioms of motion and the law of gravitational attraction were shielded from modification or replacement. Divergences that arose between calculations and observations were removed by making changes in the protective belt of auxiliary hypotheses. Thus when the motion of the moon was observed not to conform to calculations made from the Newtonian axioms, agreement was achieved by adding an auxiliary hypothesis to the protective belt. This auxiliary hypothesis attributes to the earth an asymmetrical distribution of mass (the earth bulges at the equator and is flattened at the poles). Given the fact that the earth's axis of rotation is inclined $23\frac{1}{2}$ degrees to the plane of its orbit, this asymmetrical mass distribution affects the earth's gravitational pull on the orbiting moon. And when the motion of newly discovered Uranus was observed to deviate from the orbit required by theory, another hypothesis was added to the protective belt. This hypothesis posited the existence of a planet beyond Uranus, and the hypothesis subsequently was confirmed. Lakatos insisted that the refusal to accept apparently negative evidence as counting against the core principles of a research program often *promotes* progress.

Lakatos applied the incorporation criterion to the sequence of theories generated by a research program. As long as modifications of the protective belt generate theories that incorporate their predecessors and achieve additional successes, the research program is progressive. The Newtonian program achieved a string of successes in the eighteenth and nineteenth centuries and should be appraised as "progressive" during this period.

Peter Clark suggested that the career of the kinetic theory of gases is an instance of a Lakatosian progressive scientific research program.

PETER CLARK ON THE DEVELOPMENT OF KINETIC THEORY AS A LAKATOSIAN RESEARCH PROGRAM

The atomic-kinetic research programme

The hard core and positive heuristic

The *hard core* of the atomic-kinetic research programme consisted simply of the proposition that: the behaviour and nature of substances is the *aggregate* of an enormously large number of very small and constantly moving elementary individuals subject to the laws of mechanics.

This is a very general, metaphysical statement about the constitution of matter. Testable versions of it were generated as a sequence of particular kinetic theories by the positive heuristic of the programme. This heuristic consisted of four methodological directives:

(i) Make specific assumptions as to the nature of the elementary individuals and as to their available degrees of freedom, such that all interactions among them are subject to the laws of mechanics.

(ii) Since the motions to be treated are "aggregate", although that motion is chaotic assume that for every property of that motion a mean value determined by the distribution of that property among the molecules exists.

(iii) Try to weaken or if possible eliminate the simplifying assumptions introduced to facilitate calculation once the specific assumptions of (i) and (ii) have been introduced so as to simulate, as far as possible, conditions obtaining in a "real" gas.

(iv) Use the specific assumptions introduced to investigate the internal properties of gases (e.g. the viscosity) while the macroscopic (hydrodynamic) and equilibrium properties should be derivable as *limiting cases*.

Thus each successive theory as the programme developed consisted of a particular *model* of a gas, constructed in accordance with (i) and (ii), each designed to be a closer approximation to the conditions known to obtain in a gas than the one before. Wherever anomalies arose, the refuting instance was attributed to the set of auxiliary assumptions constituting the model in question. The heuristic contained suggestions as to how to fill in, elaborate and draw consequences from each theory within the programme. It thus set out a *research policy* and gave hints and suggestions as to how it might be carried out.

＊ ＊ ＊

The empirical progress of the early kinetic programme
The elementary theory

In the original Krönig-Joule theory a very simple statistical hypothesis was adopted, viz. "the path of each molecule must be so irregular that it will defy all calculation. However, according to the laws of probability theory, one can assume a completely regular motion in place of this completely irregular one." More specifically, all the molecules can be regarded as traversing rectilinear paths with the same speed, in arrays parallel to and normal to the walls of the containing vessel. (This amounts to a particular specification of (ii) in the heuristic.) Furthermore, the molecules are *smooth* elastic spheres (specification of (i)) from which it follows that only the motion of translation is present. Impacts are perfectly elastic: From this very simple theory the equation of state (Boyle-Charles law) for an ideal gas follows, when the *vis viva* of the molecules is assumed to be proportional to the absolute temperature of the gas.

Clausius extended the simple Krönig model by attempting more realistic assumptions about molecular mechanics and a more realistic account of the state of motion of the molecules. On the basis of translatory motion alone, Clausius could not account for the observed specific heat of gases. This anomaly was attributed by him to the auxiliary hypothesis employed in Krönig's theory that the molecules were smooth elastic spheres. However, independent considerations suggested that they were not, and since not all molecular collisions could be rectilinear and central, a rotatory motion would ensue. Furthermore, the collisions could not be regarded as perfectly elastic, for if a molecule was a combination of atoms one would expect vibrations to occur among the atoms during and after impact. Thus any motion of translation alone would gradually become distributed among the other available degrees of freedom of motion. Once a steady state had been reached it would be possible to neglect the irregularities occurring as the result of these inelastic collisions and to "assume that in reference to the translatory motion, the molecules follow the common laws of elasticity." Also, since a steady state would be attained, a distribution of molecular speeds would be assumed which had a very small dispersion about a most frequent value; thus in this state all the molecules could be assumed to have a constant (the mean) speed. Most important, however, the assumption of motion in regular arrays was abandoned (in accordance with the directives of the heuristic (iii) above). Clausius adopted the method of calculating the probability of finding a molecule with speed u with a direction within an element of solid angle dw and then integrating over all directions. This innovation, together with the concept of a distribution of speeds possessing a definite spread, constituted important heuristic progress as we shall see when we consider the later theories of Clausius and Maxwell. First, however, let us examine the *novel* consequences of the elementary theory (i.e. the genuine empirical support, if any, for the theory).

Here many of the *novel* consequences were *not temporally novel;* they were novel in the sense that the theory was not specifically designed to accommodate them. Clausius's theory in effect predicted Dalton's law of partial pressures for a mixture of gases *and* Gay-Lussac's law of equivalent volumes.

Furthermore, it gave a novel qualitative explanation of the phenomenon of evaporation, and the equilibrium between liquid and gaseous phases. Thus, Clausius's first theory constituted progressive shift over the Krönig-Joule theory since it predicted a series of novel facts which did not follow from the Krönig-Joule theory.[5]

Later in the history of the kinetic theory research program, van der Waals suggested (1873) that the ideal gas law be replaced by the equation $(P + a/V^2)(V - b) = kT$, where a is a measure of intermolecular attractive forces, and b is a correction term to account for the finite volume of the molecules.

The two parameters a and b must be determined from experiment (i.e., two sets of values for P, V, T). Once this is done, however, the isotherms (constant T) can be predicted for all (P, V). The van der Waals equation, for example, gave the first theoretical determinations of the critical points (the change of state). The van der Waals isotherms and the predicted values for the critical points were confirmed in tests carried out using the empirically determined isotherms for carbon dioxide. The agreement between the van der Waals isotherms and the experimental ones was, however, by no means complete; anomalous results abounded, especially at high compression.

What is important is that the novel predictions were found to be in good agreement with experiment. Application of the heuristic of the kinetic programme had resulted in empirical growth, in this case the discovery of a *new general law,* which for the very first time allowed for the theoretical determination of some of the intractable properties of non-ideal gases. This discovery of a quite general law was arrived at, using the heuristic of the programme, quite independently of the detailed experimental investigations of Andrews into the isotherms of carbon dioxide and the numerous experimental deviations from the perfect gas law (i.e. refutations of earlier kinetic theories).

* * *

I have attempted in this section to appraise the early development of the kinetic research programme. I should now like to emphasise three important points:

First: what we have appraised *was* a research programme. That is, it was a sequence of theories, each of which consisted of specified forms of a set of hypotheses held constant (in effect regarded methodologically as unfalsifiable) and a set of auxiliary hypotheses differing in each theory in the sequence. Further, each successive theory can be seen to have been constructed in accordance with a general, overall plan.

Second: the research programme was *progressive*. That is, until the early 1880s each version of the research programme predicted a *novel fact* (moreover, as we have noticed, each version predicted a *temporally* novel fact).

Third: the overwhelming importance of the heuristic in the development of the kinetic research programme.[6]

The eighteenth-century phlogiston program, by contrast, rather quickly became a degenerating program. Phlogiston theorists interpreted combustion as a process in which the substance "phlogiston" is liberated from the burning material. In the case of the combustion of a metal, the phlogiston theory interpretation is

$$\text{metal} \xrightarrow{\Delta} \text{calx} + \text{phlogiston}$$

The theory achieved initial success by showing that combustion, respiration, and acidification are processes of the same type.

Alan Musgrave has recast this historical episode as an example of a "degenerating" Lakatosian research program.

Phlogiston before Lavoisier

The programme originated with Becher's claim that combustibles contained an "inflammable principle" which they released upon burning. Since metals upon heating turned into powdery substances like ashes, it was also claimed that they too contained the "inflammable principle", and that calcination was slow combustion. And since metallic ores, when heated with charcoal, turned into metals, it was said that in smelting ores we supply the "inflammable principle" to them. The "principle of inflammability" was thus transferred from a combustible (the charcoal) to a metal. But could it be transferred to another genuine combustible? Burning sulphur yielded vitriolic acid fumes. Stahl (who coined the term "phlogiston") fixed these fumes in potash, and heated the resulting salt (potassium sulphate) with charcoal to obtain "liver of sulphur". Since "liver of sulphur" resulted from mixing sulphur and potash, Stahl concluded that the phlogiston from the charcoal had combined with the vitriolic acid fumes to produce sulphur. Phlogiston, or the "principle of inflammability", could be transferred from one combustible to another. Lavoisier later called this the "great discovery of Stahl".

The phlogiston programme initially progressed: it gave a unified explanation of the apparently distinct phenomena of combustion and calcination, and Stahl managed to confirm a novel prediction. But there were many anomalies. Some well-known facts about combustion were not explained: why does combustion soon cease in an enclosed volume of air, and why is the volume of air reduced by it; why won't things burn at all in a vacuum? Worse still, other well-known facts seemed to refute the theory: why, if calcination is the release of phlogiston, do calxes weigh more than the original metals?

The first two anomalies were dealt with by adding auxiliary hypotheses. Phlogiston must be carried away from a combustible by the air, and a given volume of air can only absorb a certain amount of it. Hence nothing will burn in a vacuum, and combustion soon ceases in a confined space. As for the reduction in volume of the air, we need only suppose that air saturated with phlogiston ("phlogisticated air") takes up less room than ordinary air (just as cotton-wool saturated with water takes up less room than ordinary cotton-wool).

The third anomaly, the weight increase of calxes, was more troublesome, but it did not compel rejection of phlogistonism. We have here a nice example of the Duhem thesis. Phlogiston theory alone does not entail that calcination will lead to a weight loss. (By phlogiston theory I mean the emerging hard core of the phlogiston programme, the thesis that combustion and calcination involve the release of phlogiston.) To derive such a prediction we need the following additional premisses: phlogiston has weight, nothing weighty is added to the metal as it calcinates, and if something weighty is removed in a process, and nothing weighty added, then the result will weigh less than the original. The observed weight increase contradicts the conjunction of phlogiston theory with these additional premisses. One could resolve the inconsistency by rejecting the phlogiston theory: Lavoisier did just that, and inaugurated a rival programme. But one could also stay within the phlogiston programme by rejecting one or more of the additional premisses.

Incidentally, nobody at the time bothered to spell out the additional premisses as I have done. They were, as Lakatos would put it, "hidden lemmas" which first saw the light of day when they were made the target for the arrow of *modus tollens* and denied. In science, as in mathematics, theories get articulated under the impact of criticism. I have had to reconstruct this particular version of phlogiston theory from subsequent versions which were inconsistent with it.

Clearly, phlogistonists had several options for accommodating the weight increase. All of them, and a few more besides, were explored. In 1772 de Morveau said that if phlogiston is lighter than air, then removing it from a body immersed in air will cause that body to weigh more (that is, he rejected the last additional premiss). Earlier, several phlogistonists harked back to Aristotelianism and ascribed negative weight or "levity" to phlogiston (that is, they rejected the first additional premiss). This hypothesis of the levity of phlogiston has caused much levity among historians ever since; Scheele, Priestley, Cavendish and Kirwan thought it funny too, and would have nothing to do with it. The third broad option was to say that the weight of calxes was augmented by something added to them as their phlogiston was released (which involves rejecting the second additional premiss). This third option leaves it open whether phlogiston is an imponderable substance or an extremely light one; it also leaves it open exactly what is the "secondary augmenter" of the calx.

Before the phlogiston programme got underway, Boyle claimed that the weight of calxes was augmented by "fire particles" which stuck to them: "It is no wonder that, being wedged into the pores, . . . the accession of so many little bodies, that want not gravity, should, because of their multitudes, be considerable upon a balance." Several phlogistonists took over this ready-made solution to the anomaly. Earlier still, Rey claimed that the weight increase of calxes, "comes from the air, which in the vessel has been rendered denser, heavier, and in some measure adhesive, by the vehement and long continued heat of the furnace: which air mixes with the calx (frequent agitation aiding) and becomes attached to its most minute particles: not otherwise than water makes heavier sand which you throw into it and agitate, by moistening and adhering to the smallest of its grains". In the early 1770s Priestley incorporated Rey's explanation into phlogistonism: the "air" which is "precipitated" into the calx is phlogisticated or "fixed air" (together, perhaps, with some water) which has been formed by the calcination process. The discovery that calxes yielded oxygen upon reduction to metals provided yet another candidate for the secondary augmenter of the calx.

The other open question, the imponderability or otherwise of phlogiston, seemed to be settled by the next phase of the programme. If metals contain phlogiston, and vitriolic acid is sulphur deprived of phlogiston and dissolved in water, then perhaps a metal will calcinate in vitriolic acid. But metals in acids effervesced and formed salts. In 1766 Cavendish immersed zinc, iron and tin in vitriolic and hydrochloric acids, and collected the gas given off. He found that it was eleven times lighter than common air, and highly inflammable. Cavendish concluded that when metals are immersed in acids "their phlogiston flies off, without having its nature changed by the acid, and forms inflammable air". If this was correct, then metallic calxes immersed in acids should form the same salts, but no "inflammable air" should be released. This was confirmed by experiment—a brilliant success.

The identification of phlogiston with "inflammable air" (hydrogen) did pose a few puzzles. "Airs" rich in phlogiston were supposed to *inhibit* combustion, yet pure phlogiston burns! When things burn they are supposed to *release* phlogiston, so it would seem that when phlogiston burns it is released from itself! Finally, metals immersed in *concentrated* vitriolic acid yielded no "inflammable air" at all. Cavendish explained this by saying that here the "inflammable air" combines with some of the acid to produce volatile "sulphureous acid", a halfway stage between vitriolic acid and sulphur. All this made Cavendish tentative about identifying "inflammable air" with phlogiston; and in 1784 he proposed a new version of phlogistonism in which this identification was denied. Yet the appeal of his 1766 version of the theory was obvious. As Kirwan put it, phlogiston "was no longer to be regarded as a mere hypothetical substance, since it could be exhibited in an aerial form in as great a degree of purity as any other air". Priestley also came to accept the identification (along with Bergman, de Morveau and De la Métherie), and in the early 1880s provided spectacular confirmations of this version of the theory. We will come to these in due course, for we must now turn to Lavoisier and to the birth of the oxygen programme.

* * *

The discovery of the composition of water also enabled Lavoisier to account for the production of inflammable air when metals are dissolved in acids. The metals oxidise, decomposing water and releasing inflammable air, and the oxide combines with the acid to form a salt. Calxes, being already oxidised, simply form the salt. Metals dissolved in concentrated nitric or sulphuric acid yield no inflammable air because here water is not decomposed (this was a real anomaly for phlogistonists, who said that the "inflammable air" came from the metal).

Finally, Lavoisier deduced a startling new prediction: that water, traditionally used to put out fires, should, since it contains oxygen, support slow combustion and yield hydrogen. Iron filings immersed in water did indeed rust and hydrogen was collected.

Encouraged by all these successes, Lavoisier came out with his second attack on phlogistonism. Again, he displays his methodological sophistication: his chief complaint against phlogistonism is that it consists of a series of *ad hoc* devices, mutually inconsistent with each other. He concludes a beautiful summary of the history of the doctrine with the famous statement: "Chemists have made a vague principle of phlogiston which is not strictly defined, and which in consequence accommodates itself to every explanation into which it is pressed. Sometimes this principle is heavy and sometimes it is not; sometimes it is free fire and sometimes it is combined with the earthy elements; sometimes it passes through the pores of vessels and sometimes they are impenetrable to it. . . . It is a veritable Proteus which changes its form every minute."

All these things, and more, had been said of phlogiston during its long history. In its struggle to keep up with the facts, especially those predicted and discovered by oxygen theorists, confused and contradictory properties had been built into the various versions of the theory. Lavoisier was groping towards the complaint that phlogistonism represented a degenerating research programme. (He had obviously been reading Lakatos between 1777 and 1783!)[7]

Lakatos's criterion of theory replacement seems appropriate within the contexts of the historical episodes discussed above. However, Lakatos restricted its application to theories *within* a particular scientific research program. The incorporation criterion would be neither a necessary nor a sufficient condition of justified theory replacement in general. It would

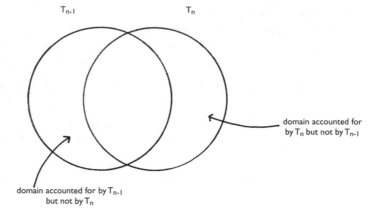

T_{n-1} T_n

domain accounted for
by T_n but not by T_{n-1}

domain accounted for by T_{n-1}
but not by T_n

Figure 18. The Partial Explanatory Overlap of Theories.

not be a necessary condition because T_n may be superior to T_{n-1} even though T_n does not account for all the successes achieved by T_{n-1}. Paul Feyerabend noted that some historical episodes display explanatory overlap. In such cases, T_n has corroborated excess content, but T_n accounts only for most—but not all—successes of T_{n-1}; see figure 18.

An example of explanatory overlap is the replacement of Ptolemy's Earth-centered solar system by Copernicus's sun-centered system. Copernicus's theory explains why Mercury and Venus are never seen at large angles from the sun (their orbits are inside the orbit of Earth); Ptolemy's theory does not. However, this explanatory gain is accompanied by an explanatory loss. The Ptolemaic theory explains why bodies dropped from a tower fall straight to its base (Earth is motionless); the Copernican theory does not.

The incorporation criterion would not be a sufficient general condition of justified theory-replacement either. Suppose theories T_1 and T_2 have been extensively confirmed for quite different ranges of application—e.g., plate tectonics theory and the kinetic theory of gases. Replacement of theory T_1, by the conjunction $(T_1 \bullet T_2)$ would qualify as "justified theory-replacement" on the incorporation criterion. $(T_1 \bullet T_2)$ accounts for all previous successes of T_1 and, in addition, has corroborated excess content. However, the conjoining of theories in this way is irrelevant to scientific progress. As Whewell emphasized, what counts is not just expansion of scope, but the manner in which this expansion is accomplished.

Moreover, the pursuit of *ad hoc* strategies may generate theories *within* a scientific research program that fulfill the requirements of the incorporation criterion without contributing to scientific progress. For example, Newton fulfilled the requirements of the criterion in developing a modified corpuscular theory of light, but it is doubtful that the revision counts as a case of justified theory replacement. To account for the rings of color produced when a slightly concave thin glass plate is pressed upon a flat glass plate, Newton hypothesized that a corpuscle of light is subject to either a "fit of easy transmission" or a "fit of easy reflexion," depending on its distance from the point of contact of the plates.[8] The revised theory accounts for a range of phenomena not explained by the original corpuscular theory, but it does so in an *ad hoc* manner. Subsequent theorists modified the corpuscular theory to account for polarization effects by postulating that a light corpuscle has "asymmetrical sides," the orientation of which determines whether the corpuscle is transmitted by certain substances. Once again, the revised theory accounts for an additional range of phenomena, but it does so in a way that proved not to be fruitful.

An analogy from evolutionary biology is pertinent here. The adequacy of a response to environmental pressure depends not only on the success of present adaptation but also on the retention of adaptability to respond to future pressures. The adequacy of a scientific theory likewise depends not only on present problem-solving effectiveness but also on potential fertility.

This suggests that Lakatos was correct to shift the focus of evaluation from individual theories to scientific research programs. That which ought to be evaluated is an entire course of research and not just the latest theory developed within the research program. To evaluate a research program is to pass judgment on its career.

The best judgments about the adaptability of a scientific research program are those rendered after the fact. Those programs (evolutionary lineages) which did undergo modification and did survive are those which initially possessed fertility (adaptability). Judgments about the present fertility of a research program (or the present adaptability of a species) are much less secure.

Lakatos emphasized that the appraisal of a scientific research program may change in the course of its development. In the case of Newtonian planetary theory, the prediction of a planet beyond Uranus was a progressive development; the subsequent prediction of a planet within the orbit of Mercury was not. The formerly progressive Newtonian program became nonprogressive in dealing with the anomalous motion of Mercury.

It also is the case that a research program that is degenerating at a given point in time subsequently may stage a comeback. Lakatos pointed to the career of William Prout's research program (1815) to show that atoms of the chemical elements are composed of multiples of hydrogen.[9] Relative to hydrogen = 1, the atomic weights of oxygen (16), nitrogen (14), sulfur (32), and carbon (12) were observed to display the required relation of multiplicity. Chlorine proved troublesome, however. Its atomic weight was determined to be 35.5. Chemists lost interest in Prout's program. It was revived in the early twentieth century, however, as applicable to the atomic numbers of the elements (the number of protons in the nucleus) rather than their atomic weights. Naturally occurring chlorine, for example, was discovered to be a mixture of two isotopes: $_{17}Cl^{35}$ and $_{17}Cl^{37}$. Each isotope has a nucleus containing 17 protons. The $_{17}Cl^{35}$ nucleus contains 18 neutrons and the $_{17}Cl^{37}$ nucleus contains 20 neutrons. Naturally occurring chlorine is comprised of 75% $_{17}Cl^{35}$ and 25% $_{17}Cl^{37}$ such that its atomic weight is 35.5. The revived (and revised) Proutian program received empirical support from X-ray diffraction studies, beginning in 1913, and became progressive once again.

Lakatos conceded that the appraisal of scientific research programs is a retrospective undertaking. It singles out those research programs that *have been* progressive. This appraisal is achieved by application of the incorporation criterion. But why appraise scientific research programs by reference to this particular criterion? To pose this question is to ask for a justificatory argument on behalf of the criterion.

There are two types of justificatory argument that may be given for an evaluative criterion. One may appeal to a philosophical theory about knowledge, and show that the criterion is required by, or at least is consistent with, that theory. Or one may appeal to historical considerations, and show that scientific evaluative practice "at its best" exemplifies the pattern prescribed by the criterion. "Logicism" and "historicism" are convenient labels for these positions.

A logicist justification of Lakatos's incorporation criterion might be given by reference to Dudley Shapere's theory of knowledge. Shapere has suggested that, given theory T, to say that "'Person P knows that T'" is to say that "P believes that T and T has been applied successfully over a period of time and no one has a specific doubt about T."[10] On this interpretation, P may know that T even if T is false. Presumably, I may know that the ideal gas law and Galileo's law of falling bodies state what is the case even though each of these laws is false. What counts are successful application and the absence of specific doubts about applying the laws to (restricted) regions of experience. Of course, one always may entertain a general doubt that experience one day may discredit a theory. But such general doubts are irrelevant to questions about knowledge.

Given this interpretation of scientific knowledge, it is reasonable to take the incorporation criterion as stipulating a sufficient condition for the *growth* of knowledge. If theories T_n and T_{n-1} satisfy the incorporation criterion, T_n has been applied successfully over time, and no specific doubts have arisen about its additional range of application, then "to know that T_n" is to have more extensive, and hence greater, knowledge than "to know that T_{n-1}." Any case of theory replacement that satisfies the incorporation criterion constitutes a gain of knowledge, given Shapere's theory of knowledge.

So far so good. The justification of the incorporation criterion—in contrast to applications of the criterion—does not invoke historical considerations. But the justification is incomplete. A logicist justification of an evaluative criterion is effective only if it can be shown that the criterion is important to the practice of science. This requires a demonstration that the criterion is exemplified in scientific evaluative practice.[11] It may be exemplified directly by certain episodes in the history of science. Or it may be exemplified indirectly as an ideal which certain episodes approximate. In either case, an effective logicist justification makes reference to the results of historical inquiry.

The historicist approach to the justification of an evaluative criterion is to demonstrate that the selected criterion is exemplified in episodes from the history of science. In addition the historicist needs to argue that the selected episodes are important. Suppose a criterion is exemplified by episodes 1–4 but is not exemplified by episode 5. The historicist needs to show that it is episodes 1–4 that are important to the advance of science and that episode 5 is not important. Episodes 1–4 constitute "science at its best"; episode 5 does not. It would seem that the historicist is forced to rely on some philosophical theory about what counts as "good science." Is there a "pure" historicist position on justification—a position that does not rely upon an antecedent philosophical standpoint?

Why not take the science of the present moment as the standard of "good science"? One then could justify those evaluative criteria whose applications led to the selection of present-day theories. To defend this approach is to presuppose a "linear progress" view of the history of science. Every stage of scientific development is taken to be superior to the preceding stage. But this linear progress view too is an underlying philosophical commitment, the defense of which involves all the problems associated with attempts to justify induction. The fact that the course of science has been uniformly progressive (if indeed it has been) does not imply that it will continue to be so. An additional difficulty for the linear progress view arises in the case of inconsistent evaluative criteria. Suppose that theory T_1 was developed to satisfy criterion c_1, theory T_2 was developed to satisfy criterion c_2, and c_1 and c_2 are inconsistent. For instance, c_1 might state that if an anomaly to a theory is discovered, the theory must be rejected or modified to account for it, and c_2 might state that an anomaly may be disregarded in the course of developing a research program. On the linear progress view, both c_1 and c_2 may be justified at a particular point in time. But a "justification" that works both for "c" and "not c" is of no value in the selection of evaluative criteria.

Lakatos developed a justificatory strategy that does not presuppose linear progress in the historical development of science. The strategy involves both logicist analysis and an appeal to historical considerations.[12]

Lakatos noted that since a philosophy of science includes criteria of theory replacement, it provides a basis for judging whether a sequence of theories is progressive. The philosophy of science thus has an interesting application to the history of science. A philosopher of science may single out those theory sequences within the history of science that qualify as progressive on his criteria. By so doing, he formulates a "rational reconstruction" of scientific progress.

Lakatos distinguished two "historical entities": the historical record of scientific developments and the philosopher's rational reconstruction of scientific progress. The historical record includes both Darwin's theory of evolution by natural selection and Lysenko's theory of inheritance; a rational reconstruction may include the former and exclude the latter.

Lakatos left unresolved the problem of the proper relationship between the history of science and its rational reconstruction. Historical episodes must be treated with respect. They must not be converted into mere illustrations of the principles of some methodology. Clearly there is a point at which a rational reconstruction is at such variance with the history of science that it no longer is a reconstruction of *science*. Musgrave pointed out that Lakatos himself is guilty of distorting history in the formulation of the "paradigm case" rational reconstruction of scientific progress—the Newtonian research program.

MUSGRAVE ON "HARD CORES" AND THE NEWTONIAN RESEARCH PROGRAM

According to Lakatos, scientists make the "hard cores" of their research programmes "irrefutable by fiat". The negative heuristic of the programme forbids them to direct the arrow of *modus tollens* towards the hard core. Instead, anomalies must be dealt with by making modifications in the protective belt, such modifications, ideally, being constructed in the light of the positive heuristic of the programme. Thus, in particular, Newton's theory (his three laws of motion and law of gravitation) was made irrefutable "by the methodological decision of its protagonists: [they decided that] anomalies must lead to changes only in the 'protective' belt of auxiliary hypotheses and initial conditions".

*   *   *

But is this explanation correct? Is it the case, to use Lakatos's own example, that Newton's theory survived for over 200 years because Newtonians took a decision not to modify or renounce it in the face of empirical difficulties? In fact, this is not the case. Throughout the history of the Newtonian research programme there were suggestions that Newton's law of gravity required modification.

In 1747 Clairault found that the amount of precession of the moon's perihelion was twice the predicted value, and suggested that Newton's law required another term involving the fourth power of the distance. But in 1749 he recalculated, and showed the modification to be unnecessary (Euler confirmed this in 1750). Then both Euler and Lagrange declared that anomalies in the secular acceleration of the moon might necessitate some modification of Newton's law. Laplace showed this to be unnecessary in 1787. Grant remarks that the publication of Laplace's *Mecanique Céleste* "forms an important landmark in the history of Physical Astronomy. The Theory of Gravitation, after being subjected to a succession of severe ordeals, from each of which it emerged in triumph, finally assumes an attitude of imposing majesty, which *repels all further question concerning the validity of its principles*". But Grant is wrong: despite its great success, the theory of gravitation had not acquired the status of a "hard core" even by 1800.

The next serious difficulty was the anomalous motion of Uranus. Several solutions to the problem were proposed. Bouvard suggested that the early, "pre-discovery observations" of Uranus, which he used to calculate the anomalous orbit, might be erroneous. But this assumed that when early astronomers observed Uranus they made errors significantly greater than their margins of error in other cases; and the anomalous motion soon reasserted itself when only modem observations were used. For the same reason, the idea that Uranus might have collided with a comet shortly before its discovery was discounted. Other suggestions were that Uranus's motion might be affected by a Cartesian cosmic fluid far out in the solar system, or by a large moon; but neither of these could account for the observed irregularities. The idea which ultimately succeeded was, of course, the postulation of an extra planet beyond Uranus. And we all know how Adams and Leverrier adopted this solution, solved the inverse perturbation problem, and successfully predicted the optical discovery of Neptune. Another apparent defeat for Newtonian astronomy had been turned into its most famous victory.

So far, this story fits well into Lakatos's pattern. The blame for the anomalous motion of Uranus was shifted, not on to Newton's theory, but on the "protective belt" of auxiliary hypotheses and initial conditions. And we can even find Leverrier, after surveying the various solutions to the problem, declaring in 1846 that to modify the law of gravitation would be "a last resort to which I would not turn until all other potential causes for the discrepancies had been investigated and rejected." But although the law of gravitation might have been something like a "hard core" for Leverrier, it was not so for others. In 1846 the Astronomer Royal, Airy, suggested that the anomaly might be accounted for by postulating that "the law or force differed slightly from that of the inverse-square of the distance". Bessel also favoured this solution at one stage. Not every Newtonian astronomer treated Newton's law as his "hard core".

It might be objected that Clairault, Euler, Lagrange, Airy, and Bessel, in contemplating modifications to Newton's law of gravity, merely showed themselves not to be true Newtonians like Laplace, Adams, or Leverrier. But this would turn Lakatos's historical claim "Newtonians rendered Newton's theory unfalsifiable by fiat" into a tautology: "Newtonians (that is, those who refused to amend Newton's laws) refused to amend Newton's laws". Definitions are not important, but it is surely better to define a Newtonian as one who contributed to the Newtonian research programme (as Clairault, Euler, Lagrange et al. surely did).

* * *

The discovery of Neptune was a staggering success, and it was hailed as such: it was "the most outstanding conceivable proof of the validity of universal gravitation", "the most noble triumph of theory which I know of", and "one of the greatest triumphs of theoretical astronomy". Nichol was so impressed with it that he formulated a methodological rule which is worth quoting in full: "there is a rule in philosophy . . . that admits of no dispute. . . . we are never entitled to challenge the universality of laws that, within our experience, have nowhere failed—*until every other mode of overcoming the difficulty has proved of no avail.* If the law of gravity must be challenged, then, the time for that is not at the beginning of our consideration of this difficulty but after we have tried every circumstance, which—the law remaining entire—might affect the manner of its working, and so have demonstrated that what is now an apparent contradiction, may not be only one of its more recondite and least obvious results." By 1850, then, Newton's law of gravity had become, for Nichol and for others, something like a "hard core".

Nichol's rule was followed in dealing with the next great difficulty which faced Newtonian astronomy, the anomalous precession of Mercury's perihelion. Leverrier tried all kinds of explanations. Perhaps the data were wrong: but recalculation using only the most precise transit observations failed to remove the anomaly. Perhaps the value for the mass of Venus was wrong: but increasing it by a tenth only removed this anomaly by introducing others. Perhaps the gravitational effects of an intra-Mercurial planet were responsible: but the search for the missing planet, which was christened "Vulcan", eventually proved unsuccessful. Perhaps the anomalous precession was produced, not by a single planet, but by several of them hitherto mistaken for sun-spots: but again, observation failed to reveal these bodies. Leverrier died in 1877, leaving the problem unsolved.

Others continued in the same vein. Perhaps the discrepancy was produced by "multitudinous small bodies individually invisible . . . mere interplanetary dust": but such a belt of dust, besides representing an *ad hoc* hypothesis, would have to be tilted at a mechanically impossible angle to do the required job. Perhaps the sun's mass is not distributed spherically (a conjecture recently revived by Dicke): but this contradicted the observed shape of the sun. Again, all these attempts fit well into Lakatos's account: they all involve modifications in the "protective belt" of auxiliary hypotheses and initial conditions. But again, these were not the *only* responses to the problem.

Some supposed that the anomaly might be explained if the gravitational force between two bodies was not merely a function of their masses and the distance between them, as Newton had said, but also of their relative velocities. Asaph Hall suggested that the exponent in Newton's law must be slightly increased. Simon Newcomb favoured this solution, and wrote, in the 1910 edition of the *Encyclopaedia Brittanica:* "the enigmatical motion of the perihelion of Mercury has not yet found any plausible explanation except on the hypothesis that the gravitation of the sun diminishes at a rate slightly greater than that of the inverse square—the most simple modification being to suppose that instead of the exponent of the distance being exactly -2 it is -2.0000001612."

Newcomb admitted that the effect of Hall's adjustment would only be detectable in the case of Mercury, so that this solution was dangerously *ad hoc*. But an *ad hoc* adjustment is still an adjustment: Newcomb had not made Newton's law of gravity "unfalsifiable by fiat"—the contrary, he thought that the Mercury perihelion had falsified it.

We surely do find this: some hypotheses have persisted over long periods despite many apparent refutations of them. But has Lakatos satisfactorily explained this feature of the history of science (which he calls continuity)? His explanation is that "hard cores" persisted because scientists took methodological decisions to retain them. And one can certainly "rationally reconstruct" history in the light of this explanation.

But to reconstruct the history of Newton's theory in this way would be to falsify it. Some Newtonians at some points did take a methodological decision to retain Newton's laws unchanged in the face of an anomaly—and some did not. The former were usually successful and the latter were not. But it is only by ignoring half of the actual history that it can be made to fit Lakatos's methodology.

* * *

To sum up: the rule "Do not blame refutations on the hard core of your research programme" was not actually followed in the "classical example of a successful research programme", is not a rule which ought to be followed, and is not necessary to explain continuity in the history of science.[13]

Lakatos conceded that rational reconstructions, such as his own retelling of the careers of Newtonian gravitation theory and Prout's program, diverge from the actual course of events. The retelling is a *rational* reconstruction, a reconstruction that conforms to the requirements of the methodology of scientific research programs. The reconstruction establishes a division between the "internal" history of science that conforms to the requirements of the methodology, and the "external" history of science that escapes reconstruction.[14]

According to Lakatos, all methodologies, including his own, are to be evaluated by appeal to the history of science. He maintained that his own methodology is superior to Baconian inductivism and Popperian methodological falsificationism. Its superiority derives from its "more inclusive" interpretation of the history of science. It renders rational more of the history of science, thereby reducing the "external" remainder.

Lakatos noted that his methodology of scientific research programs can account for the success of theories about such "nonobservables" as light waves, electrons, and quarks. For the methodology of inductivism, by contrast, the success of these theories is a mystery. For this reason, the methodology of scientific research programs is superior to inductivism. Lakatos noted, as well, that his methodology can account for the continuing success of research programs that are pursued in the face of *prima facie*

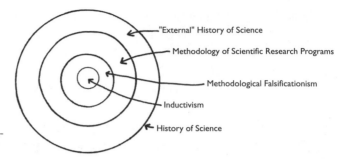

Figure 19. Lakatos on Competing Methodologies.

"External" History of Science

Methodology of Scientific Research Programs

Methodological Falsificationism

Inductivism

History of Science

falsifying evidence. For instance, scientists continued to apply Newtonian mechanics successfully in spite of the failure of its calculations to reproduce the observed orbit of Mercury. Such pursuit is consistent with the methodology of scientific research programs, but is a mystery from the standpoint of methodological falsificationism.

Lakatos's claim for the superiority of his methodology rests on its supposed "better fit" to the history of science. Thomas Kuhn criticized Lakatos for recommending a circular evaluative procedure.[15] Lakatos, following Whewell, maintained that one can write a history of science only after accepting some methodological standpoint. Lakatos also maintained that his own methodology provides the best available standpoint for the interpretation of the history of science. It seemed to Kuhn that Lakatos's justification is circular—the methodology of scientific research programs is justified upon appeal to a history of science formulated from the standpoint of this same methodology.

There is circularity here. But it remains possible, from Lakatos's standpoint, that a new methodology be developed which is superior to the methodology of scientific research programmes in the same way that this methodology is superior to falsificationism and inductivism. This new methodology would render rational additional episodes in the history of science, thereby further reducing the residual domain of "external" history of science. On Lakatos's evaluative procedure, this new methodology then should be adopted for purposes of historical reconstruction. The circularity of Lakatos's evaluative procedure is open-ended.

10. The Asymptotic Agreement of Calculations

The physicist R. E. Peierls, commenting on the reception of Einstein's theory of relativity, maintained that an acceptable new scientific theory must pass three tests.[1]

PEIERLS ON CRITERIA OF ACCEPTABILITY FOR SCIENTIFIC THEORIES

The ideas of relativity at first met with rather strong opposition both amongst physicists and amongst philosophers. The physicists met the new hypothesis in the critical spirit in which it is their business to regard any new idea, until it has passed the triple test which physics requires. It must firstly leave undisturbed the successes of earlier work and not upset the explanations of observations that had been used in support of earlier ideas. Secondly it must explain in a reasonable manner the new evidence which brought the previous ideas into doubt and which suggested the new hypothesis. And thirdly it must predict new phenomena or new relationships between different phenomena, which were not known or not clearly understood at the time when it was invented. This process took some time because relativity is important only for objects moving with a speed comparable to that of light. As such objects were not readily available the opportunities for tests were few, and in many cases the tests required very difficult observations of high precision. Since then particles moving with high velocity have become commonplace in any physics laboratory. We are no longer concerned with small corrections in the behaviour of these particles which require measurements of high precision, but the relativistic features of their motion have large effects which it is quite impossible to overlook. To put the point in its crudest form: in developing machines for physical research in which particles are accelerated to very high speeds, engineers have to incorporate devices, costing many thousands of pounds, which are required only because of the relativistic features of the particle motion. They are scarcely inclined to regard such features as a result of idle or mistaken speculation. Today no physicist who has practical knowledge of work with fast particles would question the principles of relativity.

The first test requires that a new theory "leave undisturbed" the successes of prior theories. As stated, Peierls's test is unsatisfactory. Copernicus's theory, for instance, does not "leave undisturbed" the successes achieved by Ptolemy's Earth-centered theory. Ptolemy's theory accounts for the failure to observe a stellar parallax and the fact that a body dropped from a tower hits the ground precisely at the base of the tower. Copernicus's theory, by contrast, requires that the earth revolve around the sun and rotate on its axis, and does not account for the failure to observe a stellar parallax and the observed motions of falling bodies.[2] Of course, one may argue, from the standpoint of some later time, that Ptolemy's theory did not "really explain" why parallax was not observed (prior to 1831) and why objects fall to the base of towers. But to take this approach is to risk converting Peierls's test into a truth of definition.

To enforce Peierls's first test is to disqualify any theory change that takes place across a revolutionary divide. However, if applications of the test are restricted to theories *within* scientific research programs, Peierls's position on theory replacement is similar to the position of Lakatos. Lakatos had insisted that, within a scientific research program, a successor theory T_n must account for the successes achieved by theory T_{n-1}. This "incorporation" does not require that the laws of T_{n-1} be derivable from the axioms of T_n. Nor does it require a "Whewellian consilience" in which T_n achieves incorporation by the superposition of a new set of unifying concepts upon the facts.

Indeed, perhaps all that is required for incorporation is an asymptotic agreement of calculations. On this minimal requirement, T_{n-1} is incorporated into T_n provided that calculations from T_n approach those from T_{n-1} under some limiting conditions. This type of correspondence is an important feature of Niels Bohr's theory of the hydrogen atom (1913). Bohr's theory stipulates that

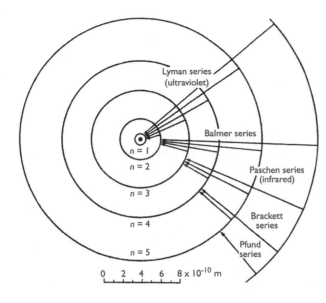

Figure 20. Bohr's Theory of the Hydrogen Atom. Gerald Holton, *Introduction to Concepts and Theories in Physical Science*, revised by Stephen G. Brush, 2nd ed. (Princeton, N.J.: Princeton University Press, 1985), p. 484.

1. The electron moves in one of a set of circular orbits around the nucleus;

2. The orbital motion of the electron obeys both Newton's second law ($F = ma$) and Coulomb's law of electrostatic attraction ($F = q_+ q_- / r^2$);

3. The orbital angular momentum is subject to a quantum condition. Its values are restricted to integral multiples of a minimum value $h/2\pi$, viz., $mv = nh/2\pi r$, where h is a constant, and n is an integer; and

4. Energy is absorbed or emitted only upon transition from one orbit to another, such that $E_i - E_f = (n_i - n_f)\ hc/\lambda$, where λ is the wavelength of the radiation.

Bohr applied this set of axioms to the emission and absorption spectra of hydrogen gas by correlating values of n and λ with the observed positions of lines in the spectra. He was able to show that transitions from orbits for which $n = 3, 4, 5 \ldots$ to $n = 2$ are in agreement with that series of spectral lines whose wavelengths were determined experimentally by Balmer. It is interesting to note that the Bohr theory provides no procedure for locating an electron, either within an orbit or between orbits during a transition.

According to Bohr's theory, the orbital energy of an electron decreases with its increasing distance from the nucleus. These decreasing energy values approach asymptotically the energy calculated from classical electromagnetic theory for a free electron no longer bound to the nucleus.

BOHR ON THE CORRESPONDENCE PRINCIPLE

The visualization of the stationary states by mechanical pictures has brought to light a far-reaching analogy between the quantum theory and the mechanical theory. This analogy was traced by investigating the conditions in the initial stages of the binding process described, where the motions corresponding to successive stationary states differ comparatively little from each other. Here it was possible to demonstrate an asymptotic agreement between spectrum and motion. This agreement establishes a quantitative relation by which the constant appearing in Balmer's formula for the hydrogen spectrum is expressed in terms of Planck's constant and the values of the charge and mass of the electron. The essential validity of this relation was clearly illustrated by the subsequent test of the predictions of the theory regarding the dependence of the spectrum on the nuclear charge. The latter result may be considered as the first step towards the fulfilment of the programme presented by the concept of the nuclear atom, to account for the relationships between the properties of the elements solely by means of the integer which represents the number of unit charges on the nucleus, the so-called "atomic number".

The demonstration of the asymptotic agreement between spectrum and motion gave rise to the formulation of the "correspondence principle", according to which the possibility of every transition process connected with emission of radiation is conditioned by the presence of a corresponding harmonic component in the motion of the atom. Not only do the frequencies of the corresponding harmonic components agree asymptotically with the values obtained from the frequency condition in the limit where the energies of the stationary states converge, but also the amplitudes of the mechanical oscillatory components give in this limit an asymptotic measure for the probabilities of the transition processes on which the intensities of the observable spectral lines depend. The correspondence principle expresses the tendency to utilize in the systematic development of the quantum theory every feature of the classical theories in a rational transcription appropriate to the fundamental contrast between the postulates and the classical theories.[3]

Bohr took the convergence of calculations established by his theory of the hydrogen atom to provide support for that theory. Ernest Hutten promoted the correspondence principle to the status of a general criterion of acceptability for theory replacement.

HUTTEN ON ASYMPTOTIC AGREEMENT

It is the correspondence principle that shows us how to construct a better and more comprehensive theory, on the basis of a simpler and narrow theory, or of a model. The principle formulates the condition which the new theory must satisfy: there must be an *asymptotic* agreement between the main formulae—of the old and of the new theory.

The new theory reduces to the old one for the special case in which the refinement introduced by the new theory may be disregarded. In other words, the new semantic system contains the calculus of the previous system as a special sub-calculus. For example, the formulae describing the orbits of the electron within the atom must become identical with those of the mechanical model for the limiting case in which the quantum of action—the concept introduced by the new theory—may be neglected. In a similar manner, the formulae for the probability distribution of a quantum-mechanical particle reduce to the equations of motion for a mechanical particle. Or, to take an example from another part of physics: the Lorentz transformation describing relative motion according to the special relativity reduces to the Galilei transformation of Newtonian mechanics, provided we deal with speeds which are small compared with that of light. This is accomplished by the procedure that certain mathematical formulae of the new theory pass, in the limiting case, into known formulae of the old theory. In this manner, the equations of quantum mechanics are given a *partial* interpretation in terms of mechanics; and the new theory, although "abstract", can be made to link up with experiment. We have reached a higher approximation, and we have achieved a closer description of nature. True that the visual model is used merely as an intermediary for the interpretation of the more advanced semantic system; however, it would be wrong to say that it is completely abandoned. We have recognised the limitations of the mechanical model; but some of the original semantic and syntactic rules are kept though others are modified or dropped altogether. Since we are not always able to formulate these rules explicitly it remains to some extent a matter of practice and skill to derive and prove theorems within the new theory. To see the restrictions imposed by the simple model, however, helps us to isolate the logical structure of the new semantic system. But it is a necessary part of theory construction that the new theory reduces to the old theory for the limiting case old theory alone has proved to be sufficient. It is in this way that we have overcome the limitations of the previous theory.

     *   *   *

The correspondence principle, then, is a rule we must follow in the construction of physical theories. When we invent a calculus for physics, we must restrict its possible interpretations; for a calculus can be interpreted, in general, in an indefinite number of ways. Since the principle states that the new theory must contain the old theory as a special case, it prescribes the conditions the new calculus must satisfy; and the possibilities for constructing new theories are curtailed in this manner. Thus, the correspondence-principle safeguards the relation to experiment. For the modern theories do not lead directly to observation sentences: the intervention of the simpler theories of classical physics is required. Bohr describes the procedure by saying that "it lies in the nature of physical observation . . . that all experience must ultimately be in terms of classical physics". For all human experience is in the medium-sized domain, and measurements are carried out by means of instruments whose functioning is described by classical physics. Classical physics is already somewhat more technical and specialised than the commonsense theory about the world which underlies ordinary language. But, as physicists, we have trained our imagination and learned to think in terms of classical physics. The correspondence principle expresses the condition that, in the last resort, we must interpret our theories of a language which is directly known to us.[4]

The "asymptotic agreement" criterion receives support from important developments in early twentieth-century physics. Calculations derived from Schrödinger's quantum theory agree with calculations derived from Newtonian mechanics in the limiting case in which the quantum of action may be neglected. And calculations derived from Einstein's special relativity theory also agree with calculations derived from Newtonian mechanics in the limiting case in which the velocity of a system is negligible with respect to the velocity of light. Many observers of science have held that this asymptotic agreement contributes to the standing of these two theories as pillars of twentieth-century physics.

Asymptotic agreement with calculations from a previously successful theory is neither a necessary nor a sufficient condition of justified theory replacement. It is not a necessary condition because successful quantitative theories often achieve improved agreement with observations over their entire range of application. Moreover, there are nonquantified theories that are superior to their predecessors.

Asymptotic agreement is not a sufficient condition of justified theory replacement either. A theorist can formulate numerous complex equations that reduce to a specified original equation under certain conditions. The asymptotic agreement of calculations is not an infallible mark of scientific progress. Consider two mathematical relations:

$$1. \ (P + a/V^2)(V - b) = kT \qquad 2. \ (P - c)(V - d/P^3) = kT$$

Calculations from both 1 and 2 agree asymptotically with those of the ideal gas law ($PV = kT$), in the limiting cases in which a and b, or c and d, approach zero. But mathematical relation #2 is an *ad hoc* construction invented solely to reduce to $PV = kT$ under specified conditions. Mathematical relation 2, however, receives support from the kinetic molecular theory of gases. Relation 1 was suggested by Johannes van der Waals (1873) to take account of intermolecular forces (the a/V^2 term) and the finite volume occupied by molecules (the b term). Van der Waals's equation provides better agreement with observed P-V-T data than does the ideal gas law at high values of pressure and temperature. Moreover, P-V-T values calculated from the van der Waals equation approach those values calculated from the ideal gas law at moderate values of these parameters.

There is an interesting case in the history of science in which a seemingly *ad hoc* mathematical relation was formulated that achieved asymptotic agreement with previously confirmed empirical laws. Max Planck (1900) proposed to treat the energy emitted from a black body[5] to be discontinuous, emerging in integral multiples of a minimum value hv, where h is a constant whose value is 6.62×10^{-27} erg.-sec. On this assumption, Planck derived an equation for black-body energy output for wavelength λ, where c is the velocity of light and k is Boltzmann's constant:

$$E_\lambda = 2\pi hc^2/\lambda^5 \exp^{(ch/\lambda kT - 1)}$$

Planck's equation is in approximate agreement with Wilhelm Wien's equation (1896) which is accurate for short wavelengths:

$$E_\lambda = a/\lambda^5 exp^{b/kT}$$

It also is in agreement with the Rayleigh-Jeans equation (1900) which is accurate for long wavelengths[6]:

$$E_\lambda = 2\pi kT/c\lambda^4$$

Planck's equation was widely regarded to be an *ad hoc* device invented to permit derivation of the Wien and Rayleigh-Jeans equations. In classical electromagnetic theory there is a continuum of possible energy values from a radiating source. Max Jammer observed that "Planck's introduction of *h* seems to be regarded at that time as an expedient mathematical device of no deeper significance, although his radiation law was repeatedly subjected to experimental test."[7]

Scientists eventually accepted the hypothesis of energy quantization as more than an *ad hoc* device. But this acceptance depended on the success of the energy quantization hypothesis in other areas, among them Einstein's theory of the photoelectric effect (1905) and Bohr's theory of the hydrogen atom (1913). Scientists initially were not impressed by the achievement of asymptotic agreement alone.

4. THE ACCEPTABILITY OF THEORIES

11. Successful Prediction and the Acceptability of Theories

Peierls's third test for justified theory replacement is that the successor theory "must predict new phenomena or new relationships between different phenomena, which were not known or not clearly understood at the time when it was invented."[1] Part of the rationale for this requirement is that it blocks the "conjunctive strategy" of theory construction. Given theory T_n and phenomenon p not accounted for by T_n, the theorist may formulate hypotheses H specifically to account for p, and then conjoin the two to form $T_{n-1} = (T_n + H)$. Since $(T_n + H)$ accounts for everything accounted for by T_n, and accounts for p in addition, the replacement of T_n by T_{n-1} satisfies the incorporation criterion. But since there are no "new" phenomena or relationships cited, and no clarification of our understanding achieved (given that H is specifically formulated to explain p), the transition $T_n \Rightarrow T_{n-1}$ does not satisfy Peierl's third test. Mere conjunction does not constitute justified theory replacement.

An example of a theory formed by conjunction is the late nineteenth-century attempt to augment Newtonian gravitation theory to account for the unidirectional motions of the planets around the sun. Kant (1755) and Laplace (1796) suggested that the planets formed from a concretion of matter contained in a series of concentric rings thrown off as a slowly rotating gaseous nebula contracted and cooled. The nebular hypothesis is consistent with Newtonian gravitation theory, and when conjoined with the Newtonian theory, provides an explanation of the unidirectionality of the planets' motions.

Despite the gain in inclusiveness, this conjunction does not qualify as an instance of justified theory replacement. The unidirectional motions of the planets around the sun was known long before the nebular hypothesis was formulated. The transition to the Newton-Laplace theory fails Peierl's requirement that *new* phenomena be accommodated by the successor theory. From Peierl's standpoint, the Newton-Laplace conjunction was unsatisfactory even before Maxwell and others showed that Laplace's nebular hypothesis is untenable. Referring to Kant's earlier version of a nebular hypothesis, George Gamow has provided a succinct account of these developments.[2]

GAMOW ON LAPLACE'S NEBULAR HYPOTHESIS

These [Kant's] views were later adopted and developed by the famous French mathematician Pierre-Simon, Marquis de Laplace, who presented them to the public in his book *Exposition du système du monde,* published in 1796. Although a great mathematician, Laplace did not attempt to give mathematical treatment to these ideas, but limited himself to a semipopular qualitative discussion of the theory.

When such a mathematical treatment was first attempted sixty years later by the English physicist Clerk Maxwell, the cosmogonical views of Kant and Laplace ran into a wall of apparently insurmountable contradiction. It was, in fact, shown that if the material concentrated at present in various planets of the solar system was distributed uniformly through the entire space now occupied by it, the distribution of matter would have been so thin that the forces of gravity would have been absolutely unable to collect it into separate planets. Thus the rings thrown out from the contracting sun would forever remain rings like the ring of Saturn, which is known to be formed by innumerable small particles running on circular orbits around this planet and showing no tendency toward "coagulation" into one solid satellite.

The only escape from this difficulty would consist in the assumption that the primordial envelope of the sun contained much more matter (at least 100 times as much) than we now find in the planets, and that most of this matter fell on the sun, leaving only about 1 per cent to form planetary bodies.

Such an assumption would lead, however, to another no less serious contradiction. Indeed if so much material, which must originally have rotated with the same speed as the planets do, had fallen on the sun, it would inevitably have communicated to it an angular velocity 5000 times larger than that which it actually has. If this were the case, the sun would spin at a rate of 7 revolutions per hour instead of at 1 revolution in approximately 4 weeks.

These considerations seemed to spell death to the Kant-Laplace views.

THE PREDICTIVIST THESIS

Instances of the "conjunctive strategy" of theory construction are quite rare. A more important reason to accept Peierl's third test is that some version of the predictivist thesis is true. A strong version of this thesis is that a theory receives more support from a fact not known or not taken into account at the time the theory is proposed than it would have received had that fact been known and taken into account. The strong predictivist thesis states that the successful prediction of a fact is more important than the mere accommodation of that same fact. Since the strong predictivist thesis makes claims about contrary-to-fact states of affairs, its status is that of an article of faith. Of course one could ask scientists "how would you have regarded the support provided by this fact if you had known about it prior to formulating your theory?" But the answers would merely be expressions of scientists' attitudes. The strong predictivist thesis will not be discussed further.

A weak version of the predictivist thesis is that successful prediction provides more support than does mere accommodation. Thus, successfully predicted evidence statement e_1 provides more support for a theory than does evidence statement e_2 that was known and taken into account when the theory was formulated. Henceforth this weak predictivist thesis will be referred to simply as the "predictivist thesis."

The most extreme form of the predictivist thesis is that *only* successful prediction counts in support of a theory. On this view, the mere accommodation of previously established facts is irrelevant. This extreme position is unacceptable. To exclude the successful accommodation of what is already known would be to eliminate important contributions to scientific knowledge.

Consider, for example, Hans Bethe's theory of solar energy production. Bethe began by constructing a theoretical model of the sun. The model stipulates a central temperature, a density gradient from center to periphery, and a specific chemical composition. Given this model, he next examined the energy outputs of the range of known nuclear reactions, and compared them to the observed rate at which the sun does radiate energy. He concluded that the only possible energy-generating mechanism for the sun is the carbon cycle in which four hydrogen nuclei are transformed into one helium nucleus.

Bethe thus succeeded in accounting for the known energy production of the sun by reference to a chain of nuclear reactions whose energy relations were known. To disqualify this achievement because no predictions were derived and confirmed, or to withhold acceptance of the theory until it was applied to other stars, would be unreasonable.

BETHE ON SOLAR ENERGY PRODUCTION

In several recent papers, the present author has been quoted for investigations on the nuclear reactions responsible for the energy production in stars. As the publication of this work which was carried out last spring has been unduly delayed, it seems worthwhile to publish a short account of the principal results.

The most important source of stellar energy appears to be the reaction cycle [1]:

$$C^{12} + H^1 = N^{13}(a), N^{13} = C^{13} + \varepsilon^+(b)$$
$$C^{13} + H^1 = N^{14}(c)$$
$$N^{14} + H^1 = O^{15}(d), O^{15} = N^{15} + \varepsilon^+(e)$$
$$N^{15} + H^1 = C^{12} + He^4(f)$$

In this cycle, four protons are combined into one α-particle (plus two positrons which will be annihilated by two electrons). The carbon and nitrogen isotopes serve as catalysts for this combination. There are no alternative reactions between protons and the nuclei $C^{12} C^{13} N^{14}$; with N^{15}, there is the alternative process, $N^{15} + H^1 = O^{16}$, but this radiative capture may be expected to be about 10,000 times less probable than the particle reaction (f). Thus practically no carbon and nitrogen will be consumed and the energy production will continue until all protons in the star are used up. At the present rate of energy production, the hydrogen content of the sun (35 percent by weight) would suffice for 3.5×10^{10} years.

The reaction cycle (1) is preferred before all other nuclear reactions. Any element *lighter* than carbon, when reacting with protons, is destroyed permanently and will not be replaced. E.g., Be^9 would react in the following way:

$Be^9 + H^1 = Li^6 + He^4$

$Li^6 + H^1 = Be^7$

$Be^7 + \varepsilon^- = Li^7$

$Li^7 + H^1 = 2He^4.$

Therefore, even if the star contained an appreciable amount of Li, Be or B when it was first formed, these elements would have been consumed in the early history of the star. This agrees with the extremely low abundance of these elements (if any) in the present stars. These considerations apply also to the heavy hydrogen isotopes H^2 and H^3.

The only abundant and very light elements are H^1 and He^4. Of these, He^4 will not react with protons at all because Li^5 is unstable, and the reaction between two protons, while possible, is rather slow and will therefore be much less important in ordinary stars than the cycle (1).

Elements heavier than nitrogen may be left out of consideration entirely because they will react more slowly with protons than carbon and nitrogen, even at temperatures much higher than those prevailing in stars. For the same reason, reactions between α-particles and other nuclei are of no importance.

Table 90-1. Energy Production in the Sun for Several Nuclear Reactions

Reaction	Average Energy Production ε (erg/g sec.)
$H^1 + H^1 = H^2 + \varepsilon^+ + f.$*	0.2
$H^2 + H^1 = He^3$	3×10^{16}
$Li^7 + H^1 = 2He^4$	4×10^{14}
$B^{10} + H^1 = c^{11} + f.$	3×10^5
$B^{11} + H^1 = 3He^4$	10^{10}
$N^{14} + H^1 = O^{15} + f.$	3
$O^{16} + H^1 = F^{17} + f.$	10-4

* "+f." means that the energy production in the reactions following the one listed, is included. E.g. the figure for the $N^{14} + H^1$ includes the complete chain (1).

To test the theory, we have calculated [Table 90-1] the energy production in the sun for several nuclear reactions, making the following assumptions:

(1) The temperature at the center of the sun is 2×10^7 degrees. This value follows from the integration of the Eddington equations with any reasonable "star model." The "point source model" with a convective core which is a very good approximation to reality gives 2.03×10^7 degrees. The same calculation gives 50.2 for the density at the center of the sun. The central temperature is probably correct to within 10 percent.

(2)The concentration of hydrogen is assumed to be 35 percent by weight, that of the other reacting element 10 percent. In the reaction chain (1), the concentration of N^{14} was assumed to be 10 percent.

(3) The ratio of the average energy production to the production at the center was calculated from the temperature-density dependence of the nuclear reaction and the temperature-density distribution in the star.

It is evident from [Table 90-1] that only the nitrogen reaction gives agreement with the observed energy production of 2 ergs/g sec. All the reactions with lighter elements would give energy productions which are too large by many orders of magnitude if they were abundant enough, whereas the next heavier element, O^{16}, already gives more than 10,000 times too small a value. In view of the extremely strong dependence on the atomic number, the agreement of the nitrogen-carbon cycle with observation is excellent.

The nitrogen-carbon reactions also explain correctly the dependence of mass on luminosity for main sequence stars. In this connection, the strong dependence of the reaction rate on temperature ($\sim T^{18}$) is important, because massive stars have much greater luminosities with only slightly higher central temperatures (e.g., Y Cygni has $T = 3.2 \times 10^7$ and $\varepsilon = 1200$ ergs/g sec.). With the assumed reaction chain, there will be no appreciable change in the abundance of elements heavier than helium during the evolution of the star but only a transmutation of hydrogen into helium. This result which is more general than the reaction chain (1) is in contrast to the commonly accepted "Aufbauhypothese."[3]

LAKATOS ON COMPARATIVE CONFIRMATION

Imre Lakatos formulated a theory of comparative confirmation in opposition to extreme predictivism. On Lakatos's view, the accommodation of previously established facts sometimes does provide evidential support for a theory. Support is forthcoming if the theory is in competition with some "touchstone theory" T_t, and the old evidence can be accommodated by T but not by T_t.[4] In this context, to say that "T_t does not accommodate e" means "T_t does not, *without modification*, explain e." Explanation, in turn, is deductive subsumption under T. Lakatos took "T explains e" to mean that "T, in conjunction with statements about relevant conditions, implies e."

In the following cases, evidence statement e, known prior to the formulation of T, nevertheless provides support for T. It does so, on Lakatos's

Evidence	Theory	Touchstone Theory
Kepler's third law	Newton's gravitation theory	Cartesian vortex theory
no pores found in the septum of the heart	Harvey's theory of circulation of the blood	Galen's theory of ebb and flow
weight calx $\geq$ weight metal	Lavoisier's oxygen theory of combustion	phlogiston theory
$C_{air} \geq C_{water}$	Maxwell's electrodynamic theory of light	Newton's corpuscular theory
anomalous motion of the perihelion of Mercury	Einstein's general theory of relativity	Newtonian mechanics
erratic blocks of granite	Agassiz's glacial theory	Werner's Neptunist theory

account, because T explains e, but T_t, which is a bona fide contender at the time T was formulated, does not explain e.

On Lakatos's view, theory assessment depends on historical considerations. Historical inquiry is required to identify the touchstone theory (or theories) against which comparison is made. A touchstone theory must be a viable competitor at the time under consideration.

Consider the case of Agassiz's glacial theory (1837). It was devised, in large measure, to explain the presence of large granite blocks in locations far removed from granite-bearing strata. Such blocks are fairly common in Northern Europe. Whether the accommodation of this old evidence by Agassiz's theory qualifies as providing support for the theory depends on whether there was a touchstone theory that could not account for the distribution of erratic granitic blocks. One might suggest Werner's Neptunist theory as a candidate for touchstone status. Since Neptunism interprets all formative geological processes, except geologically recent volcanic activity, to involve precipitation from aqueous solution or suspension, it does not have the resources to account for the erratic blocks.

A historian of science might object that Neptunism had been discredited as a general theory prior to Agassiz's formulation of the glacial theory. If so, then it was not a touchstone theory at the time. Whether Werner's theory, or Buckland's theory, or Lyell's theory had touchstone status is open to debate among historians of science.

Lakatos's theory of comparative confirmation legitimates the evidentiary status of old evidence in certain cases. The predictivist thesis nevertheless may be correct. Indeed, this thesis may receive support from developments in the history of science. The following episodes sometimes are cited on behalf of the predictivist thesis:

1. The prediction of a planet beyond Uranus by Leverrier and Adams. Presumably, Newtonian gravitation theory received greater support from the discovery of Neptune (identified by Galle after he received a prediction of its position from Leverrier) than it did from its successful derivation of the orbits of known planets. Scientists presumably gave more weight to this successful prediction than to the derivation of orbits for the known planets Mars, Jupiter, and so on.

2. Edmund Halley's prediction that a comet observed in 1682 would be observed again in 1758. Halley noted that bright comets had been recorded in 1456, 1531, 1607, and 1682. He hypothesized that the observations were of a single comet traveling around the sun in a highly elliptical orbit. Scientists took the return of the comet in 1758 to be a powerful confirmation of Newtonian gravitation theory. They did so, despite the fact that the prediction could have been made solely by hypothesizing that the sightings were of a single returning comet, and then extrapolating from past evidence of its periodicity, without recourse to Newtonian theory.

3. Mendeleyev's predictions of the properties of hitherto undiscovered elements and their compounds. (See chapter 6.)

4. S. D. Poisson's prediction that if Fresnel's wave theory of light is correct, then there should be a bright spot at the center of a shadow cast by an opaque disk illumined by a point source. Poisson was committed to the corpuscular theory of light. He believed that the "absurd" consequence he had derived from the wave theory discredited this rival to the corpuscular theory. Fresnel agreed that Poisson's derivation is correct and arranged for the relevant experiment to be performed. To the surprise of many members of the French scientific community, Poisson's prediction was confirmed (1819). This result was acknowledged to constitute important support for the wave theory of light.[5]

5. J. C. Maxwell's prediction, from the axioms of the kinetic molecular theory, that the viscosity of a gas is independent of its density at a given temperature. He noted that this is a surprising result. One would think

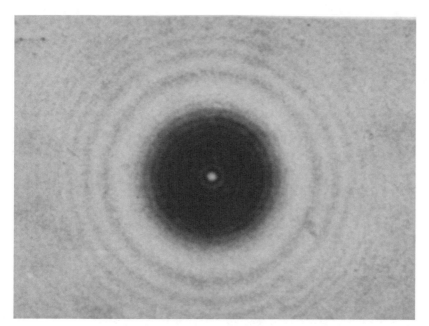

Figure 21. The Poisson Bright Spot. Vasco Ronchi, *The Nature of Light* (London: Heinemann, 1970), p. 221.

that the friction produced by layers of gas moving past one another would increase with increasing density. This is what happens in the case of liquids. Maxwell conceded (1860) that there is no experimental evidence for this consequence of kinetic molecular theory.[6] He then designed an experiment to test this unexpected consequence of the theory and reported that viscosity is indeed independent of density over a wide range of densities.

6. Einstein's prediction that the apparent position of a star is displaced as it passes behind the sun. Einstein derived this prediction from the principles of the general theory of relativity which imply a bending of light around massive objects like the sun.

7. The prediction by Fred Vine and Drummond Matthews (1963) that if Harry Hess's theory of seafloor spreading is correct, and if there has been a history of polarity reversals of the Earth's magnetism, then there should be alternating strips of normally magnetized and reverse magnetized materials along ridges formed beneath the ocean. Numerous studies of seafloor ridges in the mid-1960s confirmed this prediction.

VINE ON SEAFLOOR SPREADING

Controversy regarding continental drift has raged within the earth sciences for more than 40 years. Within the last decade it has been enlivened by the results of paleomagnetic research and exploration of the ocean basins (Bullard et al., 1965). Throughout, one of the main stumbling blocks has been the lack of a plausible mechanism to initiate and maintain drift. Recently, however, the concept of spreading of the ocean floor, as proposed by Hess (1962), has renewed for many the feasibility of drift and provided an excellent working hypothesis for the interpretation and investigation of the ocean floors. The hypothesis invokes slow convection within the upper mantle by creep processes, drift being initiated above an upwelling, and continental fragments riding passively away from such a rift on a conveyor belt of upper-mantle material: movements of the order of a few centimeters per annum are required. Thus the oceanic crust is a surface expression of the upper mantle and is considered to be derived from it, in part by partial fusion, and in part by low-temperature modification. This model, as developed by Hess (1965) and Dietz (1966), can be shown to account for many features of the ocean basins and continental margins.

It seems reasonable to assume that, if drift has occurred, some record of it should exist within the ocean basins. Heezen and Tharp (1965; 1966) have delineated north-south topographic scars on the floor of the Indian Ocean that may well be caused by the northward drift of India since Jurassic time. Wilson (1965d) has suggested that drift and ocean-floor spreading in the South Atlantic and East Pacific may be recorded in the form of fracture zones and a seismic volcanic ridges. It has also been postulated that the history of a spreading ocean floor may be recorded in terms of the permanent (remanent) magnetization of the oceanic crust.

Vine and Matthews (1963) have suggested that variations in the intensity and polarity of Earth's magnetic field may be "fossilized" in the oceanic crust and that this conditions in turn should be manifest in the resulting short-wavelength disturbances or "anomalies" in Earth's magnetic field observed at or above Earth's surface. Thus the conveyor belt can also be thought of as a tape recorder. As new oceanic crust form and cools through the Curie temperature at the center of an oceanic ridge, the permanent component of its magnetization, which predominates, will assume the ambient direction of Earth's magnetic field. A rate of spreading of a few centimeters per annum and a duration of 700,000 years for the present polarity (Cox el al., 1964b; Doell and Dalrymple, 1966) imply a central "block" of crust, a few tens of kilometers in width, in which the magnetization is uniformly and normally directed. The adjacent blocks will be of essentially reversed polarity, and the width and polarity of blocks successively more distant from the central block will depend on the reversal time scale for Earth's field in the past.

Vine and Wilson considered that the bulk of the magnetization resides in a comparatively thin layer, 1 or 2 kilometers of basaltic extrusives and intrusives, coating a main crustal layer of serpentinite (Vine and Wilson, 1965). If the frequency of extrusion and intrusion of this material is approximately normally distributed about the axis of the ridge (Loncarevic et al., 1966), all blocks other than the central block will be contaminated with younger material, possibly of opposite polarity, in which case their bulk resultant or effective magnetization will be reduced. In this way a model has been developed which the magnetization of the central block is assumed to be twice that of the others. This model derives from work on a very small but detailed survey of an area on the crest of Carlsberg Ridge in the northwest Indian Ocean (Cann and Vine, 1966).

Of the three basic assumptions of the Vine and Matthews hypothesis, field reversals (Cox et al., 1964b; and Dalrymple, 1966) and the importance of remanence (Ade-Hall, 1964) have recently become more firmly established and widely held: thus in demonstrating the efficacy of the idea one might provide virtual proof of the third assumption: ocean-floor spreading, and its various implications.[7]

8. Laplace's application of his theory of heat to explain the discrepancy between the observed velocity of sound and the calculated velocity of sound. The historian of science William Whewell noted that prominent scientists such as Lagrange and Euler had tried in vain to narrow the gap between the requirements of theory and the observed value. He declared that

> Various analytical improvements and extensions were introduced into the solution by the two great mathematicians; but none of these at all altered the formula by which the velocity of sound was expressed; and the discrepancy between calculation and observation, about one-sixth of whole, which had perplexed Newton, remained still unaccounted for.

The merit of satisfactorily explaining this discrepancy belongs to Laplace. He was the first to remark that the common law of the changes of elasticity in the air, as dependent on its compression, cannot be applied to those rapid vibrations in which sound consists, since the sudden compression produces a degree of heat which additionally increases the elasticity. The ratio of this increase depended on the experiments by which the relation of heat and air is established. Laplace, in 1816, published the theorem on which the correction depends. On applying it, the calculated velocity of sound agreed very closely with the best antecedent experiments, and was confirmed by more exact ones instituted for that purpose.[8]

Laplace reasoned that the propagation of sound involves the compression and rarefaction of the medium, and that the compression of an elastic fluid generates heat. He calculated the amount of heat produced by the propagation of sound in air and used this value to recalculate the theoretical value of the velocity of sound. LaPlace's recalculated value was in good agreement with the experimentally determined value. A puzzling discrepancy had been eliminated.

9. Einstein's application of Planck's quantum hypothesis to explain the photoelectric effect. The photoelectric effect is the setting free of electrons on the surface of a metal upon impact of a beam of light. This effect is utilized in the light meters of cameras. A light meter records the intensity of light in the field of view of a lens by measuring current values within a metal strip behind the lens.

Einstein suggested that the incident light energy is not continuous but consists of discrete "bundles" (photons). He restricted energy values of the incident photons to integral multiples of hv, where h is Planck's constant. Einstein thus extended Planck's quantum hypothesis to cover, not only the black-body oscillators that emit light, but also the emitted light itself. R. A. Millikan subsequently established (1916) that the value of h determined from experiments on the photoelectric effect is the same as the value determined from experiments on black-body radiation.

EINSTEIN ON THE PHOTOELECTRIC EFFECT

Let us begin by changing the intensity of the homogeneous violet light falling on the metal plate and note to what extent the energy of the emitted electrons depends upon the intensity of the light. Let us try to find the answer by reasoning instead of by experiment. We could argue: in the photoelectric effect a certain definite portion of the energy of radiation is transformed into energy of motion of the electrons. If we again illuminate the metal with light of the same wave-length but from a more powerful source, then the energy of the emitted electrons should be greater, since the radiation is richer in energy. We should, therefore, expect the velocity of the emitted electrons to increase if the intensity of the light increases. But experiment again contradicts our prediction. Once more we see that the laws of nature are not as we should like them to be. We have come upon one of the experiments which, contradicting our predictions, breaks the theory on which they were based. The actual experimental result is, from the point of view of the wave theory, astonishing. The observed electrons all have the same speed, the same energy, which does not change when the intensity of the light is increased.

This experimental result could not be predicted by the wave theory. Here again a new theory arises from the conflict between the old theory and experiment.

Let us be deliberately unjust to the wave theory of light, forgetting its great achievements, its splendid explanation of the bending of light around very small obstacles. With our attention focused on the photoelectric effect, let us demand from the theory an adequate explanation of this effect. Obviously, we cannot deduce from the wave theory the independence of the energy of electrons from the intensity of light by which they have been extracted from the metal plate. We shall, therefore, try another theory. We remember that Newton's corpuscular theory, explaining many of the observed phenomena of light, failed to account for the bending of light, which we are now deliberately disregarding. In Newton's time the concept of energy did not exist. Light corpuscles were, according to him, weightless; each color preserved its own substance character. Later, when the concept of energy was created and it was recognized that light carries energy, no one thought of applying these concepts to the corpuscular theory of light. Newton's theory was dead and, until our own century, its revival was not taken seriously.

To keep the principal idea of Newton's theory, we must assume that homogeneous light is composed of energy-grains and replace the old light corpuscles by light quanta, which we shall call *photons,* small portions of energy, traveling through empty space with the velocity of light. The revival of Newton's theory in this new form leads to the *quantum theory of light.* Not only matter and electric charge, but also energy of radiation has a granular structure, i.e., is built up of light quanta. In addition to quanta of matter and quanta of electricity there are also quanta of energy.

The idea of energy quanta was first introduced by Planck at the beginning of this century in order to explain some effects much more complicated than the photoelectric effect. But the photo-effect shows most clearly and simply the necessity for changing our old concepts.

It is at once evident that this quantum theory of light explains the photoelectric effect. A shower of photons is falling on a metal plate. The action between radiation and matter consists here of very many single processes in which a photon impinges on the atom and tears out an electron. These single processes are all alike and the extracted electron will have the same energy in every case. We also understand that increasing the intensity of the light means, in our new language, increasing the number of falling photons. In this case, a different number of electrons would be thrown out of the metal plate, but the energy of any single one would not change. Thus we see that this theory is in perfect agreement with observation.

What will happen if a beam of homogeneous light of a different color, say, red instead of violet, falls on the metal surface? Let us leave experiment to answer this question. The energy of the extracted electrons must be measured and compared with the energy of electrons thrown out by violet light. The energy of the electron extracted by red light turns out to be smaller than the energy of the electron extracted by violet light. This means that the energy of the light quanta is different for different colors. The photons belonging to the color red have half the energy of those belonging to the color violet. Or, more rigorously: the energy of a light quantum belonging to a homogeneous color decreases proportionally as the wave-length increases. There is an essential difference between quanta of energy and quanta of electricity. Light quanta differ for every wave-length, whereas quanta of electricity are always the same. If we were to use one of our previous analogies, we should compare light quanta to the smallest monetary quanta, differing in each country.

Let us continue to discard the wave theory of light and assume that the structure of light is granular and is formed by light quanta, that is, photons speeding through space with the velocity of light. Thus, in our new picture, light is a shower of photons, and the photon is the elementary quantum of light energy. If, however, the wave theory is discarded, the concept of a wave-length disappears. What new concept takes its place? The energy of the light quanta! Statements expressed in the terminology of the wave theory can be translated into statements of the quantum theory of radiation. For example:

Terminology of the Wave Theory	Terminology of the Quantum Theory
Homogeneous light has a definite wave-length. The wave-length of the red end of the spectrum is twice that of the violet end.	Homogeneous light contains photons of a definite energy. The energy of the photon for the red end of the spectrum is half that of the violet end.

The state of affairs can be summarized in the following way: there are phenomena which can be explained by the quantum theory but not by the wave theory. Photo-effect furnishes an example, though other phenomena of this kind are known. There are phenomena which can be explained by the wave theory but not by the quantum theory. The bending of light around obstacles is a typical example. Finally, there are phenomena, such as the rectilinear propagation of light, which can be equally well explained by the quantum and the wave theory of light.[9]

The above examples are of two types. Examples 1 through 7 are predictions of new phenomena; examples 8 and 9 display "undesigned scope," in which a theory is demonstrated to be applicable to a type of phenomena not considered at the time it was formulated.

CRITICISMS OF THE PREDICTIVIST THESIS

Critics of the emphasis placed on prediction have argued that, regardless of whether prediction is more important than accommodation, the successful prediction of new phenomena does not justify theory acceptance. For instance, Ptolemy's Earth-centered astronomical models were quite accurate in predicting eclipses and planetary positions along the zodiac. If the successful prediction of new phenomena counts for the Copernican theory, then it also counts for the Ptolemaic theory. The Copernican Sun-centered alternative is the superior theory, but this is not because of superior predictive ability.

Martin Carrier noted that both the phlogiston theory and the caloric theory have recorded predictive successes. But both theories have been discredited. Their central theoretical terms—"phlogiston" and "caloric"—fail to refer.

Nevertheless, Joseph Priestley succeeded in his prediction that the phlogiston (φ) produced when metals are dissolved in acids (1) would act like charcoal (which is rich in phlogiston) in the reduction of calces to their metals (2).[10]

1. phlogiston theory: metal (calx + φ) + acid = salt + φ ("inflammable air")
modern theory: $Zn + 2\,HCl = ZnCl_2 + H_2$

2. phlogiston theory: calx + charcoal = metal + "fixed air"
modern theory: $ZnO + C = Zn + CO_2$

Priestley confirmed this novel prediction by heating calces in the presence of pure phlogiston (hydrogen) so as to produce the corresponding metals (3).

3. phlogiston theory: calx + φ = metal
modern theory: $ZnO + H_2 = Zn + H_2O$

Priestley observed that water appeared in the experiment, but attributed it to condensation of moisture present initially in the "inflammable air" (phlogiston). Carrier declared that in this instance, "we have a theoretical prediction of an empirical regularity that was not known to science before."[11]

The second episode features a prediction derived from the caloric theory of heat. Dalton and Gay-Lussac, working independently, predicted that the rate of increase of volume with temperature at constant pressure should be the same for all gases. According to the caloric theory, heat is a substance. The increase of temperature of a gas is interpreted to be an increase in the size of the fluid atmosphere that surrounds each individual molecule of the gas (figure 22).

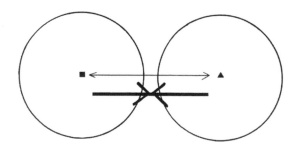

Figure 22. The Caloric Theory Model of a Gas.

On the caloric theory, the pressure exerted by a gas—its "spring"—is the result of mutually repulsive forces between atmospheres of caloric. At constant pressure, the volume of a gas increases with its temperature as more and more caloric is injected into the atmospheres around the molecules.

Dalton and Gay-Lussac argued that the rate of thermal expansion at constant pressure should be the same for all gases because the repulsive force between two atmospheres depends only on the amount of caloric involved, and the molecules are so far apart in the gas phase that the force of attraction between them is negligible. Experimental determinations confirmed this conclusion. Carrier emphasized that this is another instance of "a confirmed prediction of an empirical law that was previously unknown to the scientific community and furthermore not to be expected given the background knowledge (that is, the state of knowledge without the theory in question)."[12]

The history of planetary astronomy, the phlogiston theory, and the caloric theory should make one cautious not to overemphasize predictive success in the assessment of scientific theories. However, these examples merely show that successful prediction *alone* does not justify theory acceptance. It still may be the case that successful prediction is more important than mere accommodation.

John Worrall and Stephen Brush have insisted that accommodation was more important than successful prediction in scientists' assessments of two of the theories discussed above. Worrall has examined the response of French scientists to the successful prediction of the Poisson bright spot. Unlike some twentieth-century methodologists who cite this prediction as evidence for the predictivist thesis, Fresnel's contemporaries saw the matter differently. The official report of the prize committee of the French Academy of Science (Laplace, Biot, Poisson, Arago, Gay-Lussac, et al.) praises Fresnel's new method for observing diffraction fringes and emphasizes the successes Fresnel achieved in describing known cases of straight-edge diffraction.[13] No particular emphasis is placed on the successful novel prediction of the Poisson bright spot. Contrary to the "history and philosophy of science folklore" that draws a predictivist moral from this episode, Fresnel's contemporaries saw his achievement primarily as formulating a mathematical theory for computing diffraction patterns produced by objects of different shapes.

Brush has examined the response of the physics community to the successful prediction of the gravitational bending of light by the Sun. His conclusion is that physicists initially did not accord greater significance to this novel prediction than to the resolution of the known difficulties with Mercury's orbit.[14] Moreover, according to Brush, physicists eventually decided that the success in predicting Mercury's orbit provided the greater support for the general theory of relativity.

Peter Achinstein carried the criticism of the predictivist thesis one step further. He maintained that there are evaluative contexts in which it is irrelevant whether an evidence statement records the test of a novel prediction. If this is correct then the predictivist thesis is false.

Achinstein called attention to J. J. Thomson's criticism of Heinrich Hertz's theory of cathode rays. Hertz had hypothesized that the rays emitted from the cathode of a partially evacuated glass tube are electrically neutral. He performed experiments in 1833 in which the cathode-ray current was not deflected by an applied electric field.

Thomson pointed out that Hertz's experiment was flawed. Hertz failed to sufficiently evacuate the cathode-ray tube. There was sufficient residual air in the tube to make the air a conductor, thereby masking the effect of the applied electric field on the cathode rays. Thomson showed that cathode rays *are* deflected by an applied electric field under conditions of higher evacuation of the tube. Hertz was wrong to hold that cathode rays are electrically neutral.

The moral that Achinstein drew from this episode is that "Hertz did not claim or imply that his experimental results provide better (or weaker) support for his theory because the theory predicted them before they were obtained. Nor did Thomson in his criticism of Hertz allude to one or the other possibility. Whichever it was—whether a prediction or an explanation or neither—Hertz (Thomson was claiming) should have used a better selection procedure. This is what is criticizable in Hertz, not whether he was predicting a novel fact or examining a known one."[15]

It would seem that the predictivist thesis is false. It is not always the case that the successful prediction of an unknown fact provides greater support for a theory than the accommodation of a known fact. Nevertheless, in many instances, scientists do accord special importance to the successful prediction of unknown facts.

UNDESIGNED SCOPE

Examples 8 and 9 above are different from the first seven. They feature a display of "undesigned scope" in which an antecedently known relation, not considered in the formulation of a theory, is explained by that theory without the theory being adjusted specifically for that purpose. William Whewell held that the achievement of undesigned scope is a sufficient condition of justified theory replacement. He claimed that there is no instance in the history of science in which undesigned scope provided support for a theory that was subsequently recognized to be false.

Perhaps Whewell was overconfident. The aforementioned episodes from the histories of the phlogiston theory and the caloric theory indicate that unanticipated relationships sometimes are captured by false theories whose central terms lack reference. Does this falsify Whewell's claim? It depends on how we interpret the term "new class of facts." The reduction of calces by "pure phlogiston" is arguably not different in kind from the reduction of calces by "phlogiston-containing" charcoal. And the nonspecific nature of the thermal expansion of gases is arguably not different in kind from the phenomena for which the caloric theory initially was formulated to explain. If the reduction of calces by inflammable air and the nonspecific thermal expansion of gases are not new classes of phenomena, then they are not countercases to Whewell's claim. The issue of novelty arises for examples 8 and 9 as well. Whewell maintained that Laplace's application of his theory of heat to the propagation of sound was a case in which undesigned scope is revealed. The theory of heat was formulated independently of considerations about the discrepancy between the calculated and measured velocities of sound in air. But was the application of the theory of heat to the propagation of sound an extension of the theory to a *new class* of facts? The application was unexpected by members of the science community. It produced a "eureka" response from some physicists. But the theory of heat implied that the compression of elastic fluids is a source of heat. Since the propagation of sound involves the compression of an elastic fluid, it may be argued that Laplace's extension of his theory did not extend the scope of the theory to a new class of facts.

A theory implies a certain set of consequences, regardless of whether or not these consequences are recognized. That it required some time for scientists to recognize certain consequences of a theory should not affect the evidential status of these consequences. But if this is correct then the determination of undesigned scope rests entirely on the unexpectedness of the application.

A similar analysis may be made of Einstein's extension of Planck's quantum theory to the photoelectric effect. This extension also produced a "eureka" response from some scientists. But it is debatable whether the extension was to a new class of phenomena. At first glance, the emission of electrons from a metal foil struck by a beam of light is quite different from the emission of energy from a heated black body. But on a deeper level, it seems obvious that if emitted electromagnetic energy is quantized, then the effects of emitted electromagnetic energy are quantized as well. From this latter perspective, what Einstein achieved was recognition of what was implied by Planck's quantum theory and not an extension of the theory to a new class of phenomena.

12. Variety within Evidence

THE DESIRABILITY OF DIVERSITY

One conclusion about theory assessment that is not controversial among scientists and methodologists is that it is important to maximize diversity within the evidence that supports a theory. Consider Snel's law of the refraction of light: $\sin i / \sin r = k$, where i is the angle of incidence of a beam of light upon an interface between two media, r is the angle of refraction within the second medium, and k is a constant whose value depends on the nature of the two media.

There is general agreement that Snel's law receives greater support from experimental results for numerous media pairs and diverse angles of incidence than from evidence restricted to the air-water interface at a 30 degree angle of incidence.

With respect to variety there is a continuum from "facts of the same kind" to "facts generally regarded to expand the scope of a theory." It is the latter class that best maximizes diversity. To maximize diversity within a body of evidence is to increase the acceptability of a theory, regardless of whether the diversity is established as a result of novel prediction or mere accommodation.

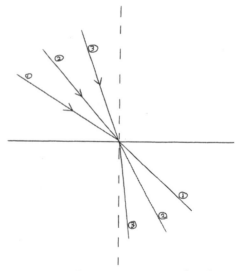

Figure 23. Refraction at Diverse Angles of Incidence.

One way in which diverse evidence for a theory is obtained is when different experimental procedures yield the same result. There are numerous episodes in the history of science in which a theory has received this type of support. Scientists have been quick to emphasize the importance of the convergence of experimental results.

The Italian chemist Stanislao Cannizzaro claimed that Avogadro's hypothesis of diatomic molecules receives important support from the coincidence of results of two types of determination of molecular formulas— volume ratios in gas-phase reactions and vapor density measurements. Consider the gas-phase reaction

nitrogen (1 vol.) + hydrogen (3 vols.) = ammonia (2 vols.)

Cannizzaro maintained that if we accept Avogadro's hypothesis (later derived from the kinetic theory of gases) that equal volumes of molecules contain equal numbers of particles, then we must conclude that nitrogen and hydrogen are diatomic gases. Consider the contrary hypothesis that nitrogen and hydrogen are monatomic gases. If this were the case, then the volume relations would have to be

nitrogen (1 vol.) + hydrogen (3 vols.) = ammonia (1 vol.)

But observation reveals that 2 volumes of ammonia are produced. Cannizzaro, following Avogadro, explained the observed volume relations by hypothesizing that nitrogen and hydrogen are diatomic molecules:

$$N_2 + 3H_2 = 2NH_3.$$

Cannizzaro pointed out that vapor density measurements support the diatomic molecule hypothesis. Consider the density measurements of gaseous compounds of hydrogen in table 5.

Each compound contains integral multiples of the weight of hydrogen present in hydrogen chloride. A standard volume of hydrogen gas contains two times this minimal weight. This gas density data also supports Avogadro's hypothesis that the molecules of hydrogen are diatomic.

5. Densities of Gaseous Hydrogen Compounds

gas	weight of 22.4 liters of gas at STP	weight of hydrogen contained	suggested formula
hydrogen chloride	36.5	1	HCl
hydrogen bromide	81	1	HBr
hydrogen cyanide	27	1	HCN
hydrogen gas	2	2	H_2
water vapor	18	2	H_2O
ammonia	17	3	NH_3
methane	16	4	CH_4

CANNIZZARO ON EVIDENCE FOR DIATOMIC MOLECULES

"Compare," I say to them, "the various quantities of the same element in the molecule of the free substance and in those of all its different compounds, and you will not be able to escape the following law: *The different quantities of the same element contained in different molecules are all whole multiples of one and the same quantity, which, always being entire, has the right to be called an atom.*"

Thus:

One molecule of free hydrogen		contains 2 of hydrogen		$= 2 \times 1$
"	of hydrochloric acid	" 1	"	$= 1 \times 1$
"	of hydrobromic acid	" 1	"	$= 1 \times 1$
"	of hydriodic acid	" 1	"	$= 1 \times 1$
"	of hydrocyanic acid	" 1	"	$= 1 \times 1$
"	of water	" 2	"	$= 2 \times 1$
"	of sulphuretted hydrogen	" 2	"	$= 2 \times 1$
"	of formic acid	" 2	"	$= 2 \times 1$
"	of ammonia	" 3	"	$= 3 \times 1$
"	of gaseous phosphuretted hydrogen	" 3	"	$= 3 \times 1$
"	of acetic acid	" 4	"	$= 4 \times 1$
"	of ethylene	" 4	"	$= 4 \times 1$
"	of alcohol	" 6	"	$= 6 \times 1$
"	of ether	" 10	"	$= 10 \times 1$

Thus all the various weights of hydrogen contained in the different molecules are integral multiples of the weight contained in the molecule of hydrochloric acid, which justifies our having taken it as common unit of the weights of the atoms and of the molecules. The atom of hydrogen is contained twice in the molecule of free hydrogen.

In the same way it is shown that the various quantities of chlorine existing in different molecules are all whole multiples of the quantity contained in the molecule of hydrochloric acid, that is, of 35.5; and that the quantities of oxygen existing in the different molecules are all whole multiples of the quantity contained in the molecule of water, that is, of 16, which quantity is half of that contained in the molecule of free oxygen, and an eighth part of that contained in the molecule of electrised oxygen (ozone).

Thus:

One molecule of free oxygen contains 32 of oxygen $= 2 \times 16$
    " of ozone   "   128   " $= 8 \times 16$
    " of water   "   16   " $= 1 \times 6$
    " of ether   "   16   " $= 1 \times 16$
    " of acetic acid   "   32   " $= 2 \times 16$
etc.

One molecule of free chlorine contains 71 of chlorine $= 2 \times 35.5$
    " of hydrochloric acid   "   35.5   " $= 1 \times 35.5$
    " of corrosive sublimate   "   71   " $= 2 \times 35.5$
    " of chloride of arsenic   "   106.5   " $= 3 \times 35.5$
    " of chloride of tin   "   142   " $= 4 \times 35.5$
etc.

In a similar way may be found the smallest quantity of each element which enters as a whole into the molecules which contain it, and to which may be given with reason the name of atom. In order, then, to find the atomic weight of each element, it is necessary first of all to know the weights of all or of the greater part of the molecules in which it is contained and their composition.

If it should appear to any one that this method of finding the weights of the molecules is too hypothetical, then let him compare the composition of equal volumes of substances in the gaseous state under the same conditions. *The various quantities of the same element contained in equal volumes either of the free element or of its compounds are all whole multiples of one and the same quantity;* that is, each element has a special numerical value by means of which and of integral coefficients the composition by weight of equal volumes of the different substances in which it is contained may be expressed. Now, since all chemical reactions take place between equal volumes, or integral multiples of them, it is possible to express all chemical reactions by means of the same numerical values and integral coefficients. The law enunciated in the form just indicated is a direct deduction from the facts: but who is not led to assume from this same law that the weights of equal volumes represent the molecular weights, although other proofs are wanting? I thus prefer to substitute in the expression of the law the word molecule instead of volume. This is advantageous for teaching, because, when the vapour densities cannot be determined, recourse is had to other means for deducing the weights of the molecules of compounds. The whole substance of my course consists in this: to prove the exactness of these latter methods by showing that, they lead to the same results as the vapour density when both kinds of method can be adopted at the same time for determining molecular weights.

The law above enunciated, called by me the law of atoms, contains in itself that of multiple proportions and that of simple relations between the volumes; which I demonstrate amply in my lecture. After this I easily succeed in explaining how, expressing by symbols the different atomic weights of the various elements, it is possible to express by means of formulae the composition of their molecules and of those of their compounds, and I pause a little to make my pupils familiar with the passage from gaseous volume to molecule, the first directly expressing the fact and the second interpreting it.[1]

At the end of the nineteenth century most physical scientists believed that atoms and molecules were real entities. There were a few dissenters, however. The usual position of the dissenters was agnosticism. The agnostic position was to accept empirical laws governing chemical combination—e.g., the laws of definite proportions, multiple proportions (Dalton), and gas-phase volume relations (Gay-Lussac)—but withhold commitment to the real existence of underlying atoms. Agnostics sought to heed the methodological lesson taught by nonreferring, but empirically successful theories, such as Ptolemaic astronomy, phlogiston theory and caloric theory. Since there was no definitive direct evidence for the existence of atoms, the proper posture is to suspend judgment.

Jean Perrin (1912) emphasized that the agreement achieved among various experimental determinations of the value of Avogadro's number N. Avogadro's number is an important physical constant. It is the number of molecules in the gram-molecular weight of any element or compound. According to the kinetic molecular theory of gases, this is the number of molecules contained in 22.4 liters of any gas under standard conditions (20° C and 1 atmosphere pressure).

The value of N may be determined experimentally by the study of numerous physical and chemical processes. N occurs in equations that describe Brownian motion, electrolysis, radioactive decay, black-body radiation, and other phenomena. Brownian motion is a zigzag movement of minute particles suspended in a gas or liquid; electrolysis is a migration of charged particles towards electrodes in liquids; radioactive decay is an emission of positively charged and negatively charged particles from the nuclei of atoms; and black-body radiation is an emission of energy from a heated cavity. Calculations of the value of N from experimental data about these processes converge on the value $N = 6.02 \times 10^{23}$ molecules/gram molecular weight. Perrin took this convergence of calculations from such very different physical and chemical processes to be decisive evidence for the truth of the atomic-molecular theory.

PERRIN ON DIVERSE DETERMINATIONS OF AVOGADRO'S NUMBER: THE AGREEMENT BETWEEN THE VARIOUS DETERMINATIONS

[A] review of various phenomena that have yielded values for the molecular magnitude enables us to draw up the following table

Phenomena observed.*	$N/10^{22}$
Viscosity of gases (kinetic theory)	62
Vertical distribution in dilute emulsions	68
Vertical distribution in concentrated emulsions	60
Brownian movement	
Displacements	64
Rotations	65
Diffusion	69
Density fluctuation in concentrated emulsions	60
Critical opalescence	75
Blueness of the sky	65
Diffusion of light in argon	69
Black body spectrum	61
Charge as microscopic particles	61(?)
Radioactivity	
Projected charges	62
Helium produced	66
Radium lost	64
Energy radiated	60

*Methods by which it may be hoped, in the future, to obtain results of great precision are given in italics.

Our wonder is aroused at the very remarkable agreement found between values derived from the consideration of such widely different phenomena. Seeing that not only is the same magnitude obtained by each method when the conditions under which it is applied are varied as much as possible, but that the numbers thus established also agree among themselves, without discrepancy, for all the methods employed, the real existence of the molecule is given probability bordering on certainty.

Yet, however strongly we may feel impelled to accept the existence of molecules and atoms, we ought always to be able to express visible reality without appealing to elements that are still invisible. And indeed it is not very difficult to do so. We have but to eliminate the constant N between the 13 equations that have been used to determine it to obtain 12 equations in which only realities directly perceptible occur. These equations express fundamental connections between the phenomena, at first sight completely independent, of gaseous viscosity, the Brownian movement, the blueness of the sky, black body spectra, and radioactivity.

For instance, by eliminating the molecular constant between the equations for black radiation and diffusion by Brownian movement, an expression is obtained that enables us to predict the rate of diffusion of spherules 1 micron in diameter in water at ordinary temperatures, if the intensity of the yellow light in the radiation issuing from the mouth of a furnace containing molten iron has been measured. Consequently the physicist who carries out observations on furnace temperatures will be in a position to check an error in the observation of the microscopic dots in emulsions. And this without the necessity of referring to molecules.

But we must not, under the pretence of gain of accuracy, make the mistake of employing molecular constants formulating laws that could not have been obtained without their aid. In so doing we should not be removing the support from a thriving plant that no longer needed it; we should be cutting the roots that nourish it and make it grow.

The atomic theory has triumphed. Its opponents, which until recently were numerous, have been convinced and have abandoned one after the other the sceptical position that was for a long time legitimate and no doubt useful. Equilibrium between the instincts towards caution and towards boldness is necessary to the slow progress of human science; the conflict between them will henceforth be waged in other realms of thought.

But in achieving this victory we see that all the definiteness and finality of the original theory has vanished. Atoms are no longer eternal indivisible entities, setting a limit to the possible by their irreducible simplicity; inconceivably minute though they be, we are beginning to see in them a vast host of new worlds. In the same way the astronomer is discovering, beyond the familiar skies, dark abysses that the light from dim star clouds lost in space takes aeons to span. The feeble light from Milky Ways immeasurably distant tells of the fiery life of a million giant stars. Nature reveals the same wide grandeur in the atom and the nebula, and each new aid to knowledge shows her vaster and more diverse, more fruitful and more unexpected, and, above all, unfathomably immense.[2]

Max Planck agreed with Perrin's emphasis on convergence. He maintained that the convergence of different kinds of experimental determinations of the value of the constant h (Planck's constant) provides important support for the theory of quantization of energy.

PLANCK ON DIVERSE DETERMINATIONS OF PLANCK'S CONSTANT

Much less simple than that of the first was the interpretation of the second universal constant of the radiation law, which, as the product of energy and time (amounting on a first calculation to $6 \cdot 55.10^{-27}$ erg. sec.). I called the elementary quantum of action. While this constant was absolutely indispensable to the attainment of a correct expression for entropy—for only with its aid could be determined the magnitude of the "elementary region" or "range" of probability, necessary for the statistical treatment of the problem—it obstinately withstood all attempts at fitting it in any suitable form, into the frame of the classical theory. So long as it could be regarded as infinitely small, that is to say for large values of energy or long periods of time, all went well; but in the general case a difficulty arose at some point or other, which became the more pronounced the weaker and the more rapid the oscillations. The failure of all attempts to bridge this gap soon placed one before the dilemma: either the quantum of action was only a fictitious magnitude, and, therefore, the entire deduction from the radiation law was illusory and a mere juggling with formulae, or there is at the bottom of this method of deriving the radiation law some true physical concept. If the latter were the case, the quantum would have to play a fundamental role in physics, heralding the advent of a new state of things, destined, perhaps, to transform completely our physical concepts which since the introduction of the infinitesimal calculus by Leibniz and Newton have been founded upon the assumption of the continuity of all causal chains of events.

Experience has decided for the second alternative. But that the decision should come so soon and so unhesitatingly was due not to the examination of the law of distribution of the energy of heat radiation, still less to my special deduction of this law, but to the steady progress of the work of those investigators who have applied the concept of the quantum of action to their researches.

The first advance in this field was made by A. Einstein, who on the one hand pointed out that the introduction of the quanta of energy associated with the quantum of action seemed capable of explaining readily a series of remarkable properties of light action discovered experimentally, such as Stokes's rule, the emission of electrons, and the ionization of gases, and on the other hand, by the identification of the expression for the energy of a system of resonators with the energy of a solid body, derived a formula for the specific heat of solid bodies which on the whole represented it correctly as a function of temperature, more especially exhibiting its decrease with falling temperature. A number of questions were thus thrown out in different directions, of which the accurate and many-sided investigations yielded in the course of time much valuable material. It is not my task today to give an even approximately complete report of the successful work achieved in this field; suffice it to give the most important and characteristic phase of the progress of the new doctrine.

First, as to thermal and chemical processes. With regard to specific heat of solid bodies, Einstein's view, which rests on the assumption of a single free period of the atoms, was extended by M. Born and Th. von Karman to the case which corresponds better to reality, viz. that of several free periods; while P. Debye, by a bold simplification of the assumptions as to the nature of the free periods, succeeded in developing a comparatively simple formula for the specific heat of solid bodies which excellently represents its values, especially those for low temperatures obtained by W. Nernst and his pupils, and which, moreover, is compatible with the elastic and optical properties of such bodies. But the influence of the quanta asserts itself also in the case of the specific heat of gases. At the very outset it was pointed out by W. Nernst that to the energy quantum of vibration must correspond an energy quantum of rotation, and it was therefore to be expected that the rotational energy of gas molecules would also vanish at low temperatures. This conclusion was confirmed by measurements, due to A. Eucken, of the specific heat of hydrogen; and if the calculations of A. Einstein and O. Stern, P. Ehrenfest, and others have not as yet yielded completely satisfactory agreement, this no doubt is due to our imperfect knowledge of the structure of the hydrogen atom. That "quantized" rotations of gas molecules (i.e. satisfying the quantum condition) do actually occur in nature can no longer be doubted, thanks to the work on absorption bands in the infra-red of N. Bjerrum, E. v. Bahr, H. Rubens and G. Hettner, and others, although a completely exhaustive explanation of their remarkable rotation spectra is still outstanding.

Since all affinity properties of a substance are ultimately determined by its entropy, the quantic calculation of entropy also gives access to all problems of chemical affinity. The absolute value of the entropy of a gas is characterized by Nernst's chemical constant, which was calculated by O. Sackur by a straightforward combinatorial process similar to that applied to the case of the oscillators, while H. Tetrode, holding more closely to experimental data, determined by a consideration of the process of vaporization, the difference of entropy between a substance and its vapour.

While the cases thus far considered have dealt with the states of thermodynamical equilibrium, for which the measurements could yield only statistical averages for large numbers of particles and for comparatively long periods of time, the observation of the collisions of electrons leads directly to the dynamic details of the processes in question. Therefore the determination, carried out by J. Franck and G. Hertz, of the so-called resonance potential or the critical velocity which an electron impinging upon a neutral atom must have in order to cause it to emit a quantum of light, provides a most direct method for the measurement of the quantum of action. Similar methods leading to perfectly consistent results can also be developed for the excitation of the characteristic X-ray radiation discovered by C. G. Barkla, as can be judged from the experiments of D. L. Webster, E. Wagner, and others.

The inverse of the process of producing light quanta by the impact of electrons is the emission of electrons on exposure to light-rays, or X-rays, and here, too, the energy quanta following from the action quantum and the vibration period play a characteristic role, as was early recognized from the striking fact that the velocity of the emitted electrons depends not only on the intensity but on the colour of the impinging light. But quantitatively also the relations to the light quantum, pointed out by Einstein, have proved successful in every direction, as was shown especially by R. A. Millikan, by measurements of the velocities of emission of electrons, while the importance of the light quantum in inducing photochemical reactions was disclosed by E. Warburg.

Although the results I have hitherto quoted from the most diverse chapters of physics, taken in their totality, form an overwhelming proof of the existence of the quantum of action, the quantum hypothesis received its strongest support from the theory of the structure of atoms (Quantum Theory of Spectra) proposed and developed by Niels Bohr. For it was the lot of this theory to find the long-sought key to the gates of the wonderland of spectroscopy which since the discovery of spectrum analysis up to our days had stubbornly refused to yield. And the way once clear, a stream of new knowledge poured in a sudden flood, not only over this entire field but into the adjacent territories of physics and chemistry. Its first brilliant success was the derivation of Balmer's formula for the spectrum series of hydrogen and helium, together with the reduction of the universal constant of Rydberg to known magnitudes; and even the small differences of the Rydberg constant for these two gases appeared as a necessary consequence of the slight wobbling of the massive atomic nucleus (accompanying the motion of electrons around it). As a sequel came the investigation of other series in the visual and especially the X-ray spectrum aided by Ritz's resourceful combination principle, which only now was recognized in its fundamental significance.

But whoever may have still felt inclined, even in the face of this almost overwhelming agreement—all the more convincing, in view of the extreme accuracy of spectroscopic measurements—to believe it to be a coincidence, must have been compelled to give up his last doubt when A. Sommerfeld deduced, by a logical extension of the laws of the distribution of quanta in systems with several degrees of freedom, and by a consideration of the variability of inert mass required by the principle of relativity, that magic formula before which the spectra of both hydrogen and helium revealed the mystery of their "fine structure" as far as this could be disclosed by the most delicate measurements possible up to the present, those of F. Paschen—a success equal to the famous discovery of the planet Neptune, the presence and orbit of which were calculated by Leverrier [and Adams] before man ever set eyes upon it. Progressing along the same road, P. Epstein achieved a complete explanation of the Stark effect of the electrical splitting of spectral lines, P. Debye obtained a simple interpretation of the K-series of the X-ray spectrum investigated by Manne Siegbahn, and then followed a long series of further researches which illuminated with greater or less success the dark secret of atomic structure.

After all these results, for the complete exposition of which many famous names would here have to be mentioned, there must remain for an observer, who does not choose to pass over the facts, no other conclusion than that the quantum of action, which in every one of the many and most diverse processes has always the same value, namely $6 \cdot 52.10^{-27}$ erg. sec., deserves to be definitely incorporated into the system of the universal physical constants. It must certainly appear a strange coincidence that at just the same time as the idea of general relativity arose and scored its first great successes, nature revealed, precisely in a place where it was the least to be expected, an absolute and strictly unalterable unit, by means of which the amount of action contained in a space-time element can be expressed by a perfectly definite number, and thus is deprived of its former relative character.[3]

Perrin and Planck advanced strong arguments for the importance of the convergence of diverse experimental determinations on a single value. A demonstration of convergence provides a warrant for a *type* of theory. In the case of determinations of Avogadro's number, for instance, the warrant is for theories that postulate a particulate microstructure that is responsible for macroscopic phenomena. Numerous individual theories about the structure and properties of the microstructure may be formulated that satisfy the convergence criterion.

13. Other Factors in Theory Evaluation

Several criteria of theory replacement have been discussed thus far. Among these criteria are incorporation, successful prediction, asymptotic agreement of calculations, and undesigned scope. There is more to theory evaluation than the application of criteria however. Thomas Kuhn has insisted that "the choices scientists make between competing theories depend not only on shared criteria—those my critics call objective—but also on idiosyncratic factors dependent on individual biography and personality."[1]

Bacon's Idols

Kuhn did not disparage the idiosyncratic factors that influence theory choice. Francis Bacon, who first called attention to such factors in the *Novum Organum* of 1620, believed them to be obstacles in the path of a correct interpretation of nature. Bacon referred to these factors as "idols." They are predispositions and prejudices that prevent the scientist from achieving an objective encounter with the world. Bacon insisted that idols be set aside so that the scientist confront nature as did Adam upon his creation. He declared that idols "must be renounced and put away with a fixed and solemn determination, and the understanding thoroughly fixed and cleansed; the entrance to the kingdom of man, founded on the sciences, being not much other than the entrance into the kingdom of heaven, whereinto none may enter except as a little child."[2]

Bacon divided idols into four classes. "Idols of the tribe" are ingrained in human nature itself. We tend to see more regularity in phenomena than really is present. We tend to overestimate the importance of supporting evidence and downplay negative evidence. "Idols of the cave" are individual preferences, determined in part by our educational experiences. Some people develop a propensity to generalize, even when this requires that they gloss over significant differences. Other people become so preoccupied by the nuances that distinguish one instance from another that they miss that which is common among them. "Idols of the marketplace" arise from social association. Bacon maintained that the terms that scientists require for the correct interpretation of nature are debased by their appropriation by the vulgar. "Idols of the theater" are the received dogmas of the philosophical schools. These idols are especially pernicious. Bacon was distressed in particular by investigators who view the universe through the distorting lenses of Aristotelian categories and assumptions.

Duhem on National Preferences

Pierre Duhem subsequently extended Bacon's analysis to include national preferences for certain types of theories. Surveying late nineteenth-century physics, he observed that Continental scientists preferred abstract theories whereas English scientists preferred theories that invoke mechanical models. Duhem singled out Lord Kelvin (William Thomson) as the leader of the English school. Kelvin had declared, "my object is to show how to make a mechanical model which shall fulfill the conditions required in the physical phenomena that we are considering. . . . It seems to me that the test of 'Do we or do we not understand a particular subject in physics?' is 'Can we make a mechanical model of it?'"[3]

Kelvin complained that abstract theories such as the dynamical theory of heat and the wave theory of light are obscure because, thus far, no single acceptable mechanical model has been provided either for heat transfer or for light propagation. Indeed, Kelvin was forced to employ different, incompatible models to represent light and the aethereal medium through which it moves, depending on whether refraction, polarization or some other phenomenon is under consideration.

Kelvin conceded, however, that the above-mentioned theories of heat and light have received experimental confirmation and are useful for purposes of prediction. Kelvin thus distinguished "acceptable" theories like these theories of heat and light, and "complete" theories which are both acceptable and possess associated mechanical models.

Kelvin held the seemingly paradoxical position that although models are required for full understanding, they are not explanations of that which is modeled. Models do not suffice to explain the processes modeled because there are good reasons to believe that model-objects such as rigid spheres, spiral springs, and cords are not fully exemplified in the real world. For instance, we have good reasons to believe that molecules in reality are not rigid spheres connected by springs. We may picture the absorption of energy by real molecules in terms of vibrations of spring-linked rigid spheres, but this picture does not explain heat absorption.

Duhem was highly critical of the English preoccupation with models. He cited with approval Kelvin's admission that this preoccupation does not serve the purposes of explanation.[4] Duhem maintained, in addition, that the history of science reveals that the consideration of models played a negligible role in the development of successful new theories.

But if the construction of models provides neither explanation nor heuristic benefit, what else remains? The English preoccupation with models is one of those Kuhnian "idiosyncratic factors" that influence which theories are entertained and accepted. Duhem included this preoccupation among the Baconian idols that should be exorcised.

Mass Extinction and Meteoric Impact

Kuhn's thesis about "idiosyncratic factors" receives additional support from the reactions of geologists and paleontologists to the Alvarez theory (1980) that a meteor impact was responsible for the mass extinctions of dinosaurs and other species at the end of the Cretaceous period. The theory was formulated to account for the discovery near Gubbio, Italy, of a high concentration of iridium in a clay layer that separates Cretaceous period strata from Tertiary period strata. Iridium is a platinum-group element that is present in significant amounts in the earth's core, but not in its crust. Iridium was known to be present, however, in many meteorites.

Dinosaurs and conifers were dominant in the Late Cretaceous period, but disappeared from the scene at the beginning of the Tertiary period.

Prior to 1980, other meteor-impact hypotheses had been formulated to account for the Late Cretaceous extinctions. There also were extinction hypotheses based on volcanic activity. The Alvarez theory differed from its predecessors because it implied several specific testable consequences.

If the earth was struck by a sizeable iridium-rich meteor at the end of the Cretaceous period, then the resultant dust cloud should have distributed iridium widely on the earth's surface. Thus there should be high iridium concentrations in the K/T (Cretaceous/Tertiary) boundary layer in various locations within the earth's crust. During the 1980s, this consequence of the Alvarez theory was confirmed.

A second consequence of the Alvarez theory is that the impact of the meteor created a sizeable crater on the earth's surface. Alvarez estimated the crater to have a diameter of about 200 kilometers. One plausible candidate was located off the coast of Mexico's Yucatan Peninsula in 1990. The crater is the right size and age and contains the required concentration of iridium.

Additional evidence was uncovered in the late 1980s and 1990s. Microscopic examination of quartz grains found in the K/T boundary layer were found to display shock lamellae, dislocations of the crystal lattice consistent with, and indicative of, meteoric impact.

Small clay-covered glass spheres were found in K/T strata. A likely origin of these spheres is a meteoric impact. The impact presumably threw molten glassy drops into the atmosphere. The drops subsequently became encased in the clay of the K/T strata.

Analysis of samples of a K/T stratum in Alberta revealed the same ratio of carbon to iridium that is found in type C2 chondritic meteorites. Moreover, the C^{12}/C^{14} ratio in the diamonds from this source corresponds to the ratio found in interstellar dust, and is not the ratio found in normal terrestrial samples. Amino acids, some not found elsewhere on earth, and some extremely rare on earth, were discovered in K/T strata. These amino acids are abundant in carbon-containing chondritic meteorites.[5]

Many geologists and paleontologists have remained skeptical about the Alvarez theory. Initially there were doubts about the postulated meteoric impact. But evidence that supported such an impact accumulated throughout the 1980s and 1990s. Some scientists then accepted the causal hypothesis that attributes mass extinctions to the effects of meteoric impact. Other scientists conceded the reality of a meteoric impact (or impacts), but remained skeptical about a causal link between impact and species extinction.

William Glen has suggested a number of reasons for the continuing skepticism about the Alvarez theory. These reasons include the ideological commitments and specific disciplinary preoccupations of the scientists involved.

GLEN ON SCIENTISTS' RESPONSES TO THE
ALVAREZ IMPACT HYPOTHESES

The generally skeptical, even poor, reception that the impact hypothesis received initially might have been predicted, for any of several reasons: such an instantaneous catastrophe flew in the face of earth science's reigning philosophy of uniformitarianism; it was based on a form of evidence alien to the community charged with its appraisal; it invoked a causal mechanism that was unlikely in terms of canonical knowledge; and it was proffered mainly by specialists outside of earth science paleobiology.

It became clear early on that *irrespective of which mass-extinction hypothesis a scientist chose, the chosen hypothesis became the strongest predictor of how the scientist would select and apply standards in assessing evidence bearing on all such hypotheses. Somewhat weaker correlations appeared between disciplinary specialty and choice of hypothesis, and between disciplinary specialty and how evidence was assessed.* Such correlations varied with level of subspecialty: the strongest correlations appeared in the most restricted subspecialties, the weakest in the broadest categories of science practice. And the gestalt or mindset seemingly engendered by subscription to a particular extinction hypothesis overrode the intellectual predispositions attributable to disciplinary specialty.

The vast majority of paleontologists rejected impact theory straightaway, and most who later came to believe in one or more impact events still deny that impact(s) are the main extinction cause. Many paleontologists viewed the duration, severity, and other aspects of the K/T mass extinction in terms of the fate of their own fossil groups at the extinction boundary; specialists in severely affected taxa generally supported the idea that an impact caused the mass extinction. The subspecialty communities within paleontology differed markedly in their opinions on the cause of mass extinction: vertebrate paleontologists, for example, were generally opposed to impact-as-extinction-cause; in contrast, micropaleontologists, especially those treating planktonic calcareous forms, were most often oppositely inclined.

Most cosmo- and geochemists, planetary geologists, impacting specialists, and those concerned with Earth-crossing comets and asteroids embraced impact theory in its entirety. Familiarity with impacting studies appeared to foster sympathy for the impact hypothesis; most volcanologists (volcanic specialists), however, did not accept the volcanism hypothesis—which was not conceived by volcanologists—but neither did any unusual proportion of volcanologists favor the alternative impact hypothesis.

In all cases examined save one, published authors of alternatives to the impact hypothesis of mass extinction, as well as published supporters of these alternatives, opposed the impact hypothesis at its advent. Irrespective of their discipline, scientists rarely failed to embrace one of the various mass-extinction hypotheses, however poorly informed they were.

The use of obsolete data and/or the omission of contrary evidence almost always punctuated published and oral arguments, and opposing views were never treated at equal length. The gestalts (mindsets) or cognitive frames of members of the opposing theoretical camps seemingly precluded mutually congruent viewpoints on any of the important debated issues, or even the assessment of evidence. So deaf did they appear to each other's arguments that Kuhn's (1970) view that such adversaries suffer an "incommensurability of viewpoint" seems understated.

Impactors and volcanists routinely invoked different standards of appraisal and weighted the same evidence differently. The application of standards and the weighting of evidence varied widely, even within the same theoretical camp. The impactors argued mainly from canonical standards and claimed that their hypothesis—backed by empirical evidence—facilitated clear predictions with implicit directions for testing. The prediction of impact-generated, global, ballistic transport of impact products raised expectations for many that impact evidence in addition to iridium would be found at K/T boundary sites, and these expectations furnished great impetus for the rapid formation of diversely comprised research teams.

Impact's opponents focused on the great range of variation in the character of the boundary interval around the world, emphasizing that the differences found from place to place betoken a cause that was neither global nor instantaneous. Demonstrating these differences became a central mission of the opponents of impact. Unlike the impactors, who in the main advanced orthodox evidence for their radical hypothesis, impact's active opponents—mainly the volcanists—sought to undermine a wide range of suppositions and vagaries that had lain long hidden in established principles and methods; the volcanists impugned orthodox standards, and thus prompted much research at unprecedented levels of refinement. The call for higher resolution required, for both camps, the development of new, or the acquisition of theretofore unacquired methods, techniques, and instruments; such needs were often fulfilled through the formation of appropriately composed collaborative teams. These great new efforts spawned proposals to various funding agencies for research grants. When such proposals were refused, undercurrents of opinion about the partisan bent of this or that funding agency spread within opposed theoretical camps.

During the mid-1980's the postulate of stepped or multiple extinctions near the K/T boundary evolved, through a series of recognizable stages. The evidence for stepped extinctions was initially viewed as an anomaly within the broader terms of the single-impact hypothesis and, as such, was more or less dismissed early on by the impactors. But as supporting data accrued, it became an increasingly serious problem, and stepped extinctions approached the status of a normative assumption. Finally, with still further affirmative evidence at hand, stepped extinctions and all that they came to imply, in terms of impugning single-impact theory, evolved into a standard of appraisal that forced a reassessment of the nature of extinctions at the K/T boundary. *At that point the newly born standard of multiple extinction steps—which had begun life as an anomaly in terms of the single-impact hypothesis—evolved into the primary forcing function that drove the reformulation of the single-impact hypothesis of multiple impacts spread over enough time to accommodate the extinction steps.*

In all cases, the leadership of the various factions engaged in the debates was clearly in the hands of only one or a very few senior leaders, who exercised all but magisterial authority. Such doyens were most frequently sought out, by both their own communities and the media, for their opinions on the debated issues; this was reflected in both the publications of science and the media of the public. The rapid pace of the mass-extinction debates, with which only few could keep abreast, seemed to add to a general reliance on the magister. Such two-step communication, from the world to the magister and then from the magister to the world, has been documented in other studies of conflicted ideas.

Definitive closure has not been reached on any of the many issues entrained in these debates, but the vast majority of earth scientists are now convinced of at least one impact at the Cretaceous/Tertiary boundary, and a substantial number are inclined to think in terms of multiple impacts, either instantaneous or spread over 1 to 3 million years; far from all who subscribe to impact(s)—especially among the paleontologists—view impact(s) as the chief cause of the mass extinction(s).[6]

Uniformitarianism is a directive principle championed by Charles Lyell (1830) and endorsed subsequently by a majority of historical geologists and evolutionary biologists. The investigator is directed to inter-pret the past by reference to processes that operate uniformly across time. Lyell recommended an extreme version of this principle, maintaining not only that the same type of processes—formative and erosive—operate throughout geological history, but also that they do so with roughly the same intensity.

A commitment to a uniformitarian perspective on the part of geologists and paleontologists was largely responsible for the initial negative reaction to the Alvarez meteoric impact theory. The Gould-Eldredge theory of punctuated equilibria (1972) earlier had received a similar negative appraisal from scientists committed to uniformitarianism.

HOLTON'S THEMATIC PRINCIPLES

Gerald Holton included directive principles such as uniformitarianism among those "themata" that influence theory choice in science. Themata express the commitments of scientists to preferred modes of inquiry, explanation, and theory evaluation.[7]

Directive Principles

Other directive principles cited by Holton are "seek quantities that are conserved, maximized, or minimized" and "seek to interpret macroscopic phenomena by reference to theories of microstructure." In addition, one might add Laplace's short-lived directive to "interpret all phenomena to be the results of central-force-field interactions among constituent particles."

Directive principles give preference to certain types of theories. Since scientists, both in their pronouncements and their actions, display commitments to such principles, a descriptive philosophy of science must acknowledge the importance of directive principles in theory appraisal.

Evaluative Standards

Holton's collection of themata includes not only directive principles, but also evaluative standards, explanatory goals, ontological assumptions, and high-level substantive hypotheses. Evaluative standards—incorporation, successful prediction, asymptotic agreement, undesigned scope, and so on—have been discussed earlier in this volume. It is incontrovertible that scientists have applied these standards within the history of science.

Explanatory Goals

Explanatory goals establish requirements for the completeness of scientific explanations. In this way they augment the evaluative standards that merely mark off those interpretations that are acceptable. Since the time of Aristotle, scientists have elevated preferred types of explanation to the status of explanatory goals.

Aristotle on Teleological Explanations

Aristotle insisted that to fully explain a natural process it is necessary to specify its telos. This is accomplished by constructing a teleological explanation of the form "X occurred *in order that* Y be achieved." For example, "that chameleon's skin became grey as it moved from leaf to bark *in order that* it be undetected by its predators," and "that stone fell when released *in order to* reach its natural place (the center of the universe)."

Aristotle maintained that teleological explanations ordinarily do not presuppose conscious deliberation and choice. The exception is in cases of human volitional activity. But teleological interpretations do presuppose that a future state of affairs (Y) determines the way in which a present state of affairs (X) unfolds. It is primarily for this reason that such interpretations have been controversial in the subsequent history of science.

Aristotle castigated the Pythagoreans for their exclusive use of formal explanations and the atomists for their exclusive use of mechanistic explanations. According to Aristotle, Pythagoreans and atomists fail to recognize what is required for a complete scientific explanation. He appealed to an explanatory goal in order to discredit interpretations that failed to satisfy it.

Galileo on the Language of Nature

Centuries later Galileo sought to counteract the widespread preoccupation with teleological accounts by emphasizing the indispensable role of mathematics in gaining knowledge of the world. His influential statement of this explanatory ideal is this: "Philosophy is written in this grand book—I mean the universe—which stands continually open to our gaze, but it cannot be understood unless one first learns to comprehend the language and interpret the characters in which it is written. It is written in the language of mathematics, and its characters are triangles, circles, and other geometrical figures, without which it is humanly impossible to understand a single word of it."[8]

Galileo's ideal of a mathematical theory of the universe serves as both a heuristic principle that directs the scientist to formulate a certain type of theory and a principle that disqualifies (or at least subordinates) other types of theories.

Einstein's Commitment to an Ideal of Determinism

Einstein subscribed to the Galilean ideal. He also was committed to an ideal of explanatory completeness that requires a deterministic account of individual events. This ideal takes statistical generalizations about ensembles of particles (or processes) to be incomplete. Ultimate explanations, Einstein insisted, specify what takes place in individual cases. Commenting on the Copenhagen interpretation of quantum mechanics, Einstein wrote the following.[9]

EINSTEIN ON THE EXPLANATION OF INDIVIDUAL EVENTS

What does not satisfy me in that theory, from the point of principle, is its attitude towards that which appears to me to be the programmatic aim of all physics: the complete description of any (individual) real situation (as it supposedly exists irrespective of any act of observation or substantiation). Whenever the positivistically inclined modern physicist hears such a formulation his reaction is that of a pitying smile. He says to himself: "there we have the naked formulation of a metaphysical prejudice, empty of content, a prejudice, moreover, the conquest of which constitutes the major epistemological achievement of physicists within the last quarter-century. Has any man ever perceived a 'real physical situation'? How is it possible that a reasonable person could today still believe that he can refute our essential knowledge and understanding by drawing up such a bloodless ghost?" Patience! The above laconic characterization was not meant to convince anyone; it was merely to indicate the point of view around which the following elementary considerations freely group themselves. In doing this I shall proceed as follows: I shall first of all show in simple special cases what seems essential to me, and then I shall make a few remarks about some more general ideas which are involved.

We consider as a physical system, in the first instance, a radioactive atom of definite average decay time, which is practically exactly localized at a point of the co-ordinate system. The radioactive process consists in the emission of a (comparatively light) particle. For the sake of simplicity we neglect the motion of the residual atom after the disintegration-process. Then it is possible for us, following Gamow, to replace the rest of the atom by a space of atomic order of magnitude, surrounded by a closed potential energy barrier which, at a time $t = 0$, encloses the particle to be emitted. The radioactive process thus schematized is then, as is well known, to be described—in the sense of elementary quantum mechanics—by a ψ-function in three dimensions, which at the time $t = 0$ is different from zero only inside of the barrier, but which, for positive times, expands into the outer space. This ψ-function yields the probability that the particle, at some chosen instant, is actually in a chosen part of space (i.e., is actually found there by a measurement of position). On the other hand, the ψ-function does not imply any assertion *concerning the time instant of the disintegration* of the radioactive atom.

Now we raise the question: Can this theoretical description be taken as the *complete* description of the disintegration of a single individual atom? The immediately plausible answer is: No. For one is, first of all, inclined to assume that the individual atom decays at a definite time; however, such a definite time-value is not implied in the description by the ψ-function. If, therefore, the individual atom has a definite disintegration time, then as regards the individual atom its description by means of the ψ-function must be interpreted as an incomplete description. In this case the ψ-function is to be taken as the description, not of a singular system, but of an ideal ensemble of systems. In this case one is driven to the conviction that a complete description of a single system should, after all, be possible; but for such complete description there is no room in the conceptual world of statistical quantum theory.

* * *

Within the framework of statistical quantum theory there is no such thing as a complete description of the individual system. More cautiously it might be put as follows: The attempt to conceive the quantum-theoretical description as the complete description of the individual systems leads to unnatural theoretical interpretations, which become immediately unnecessary if one accepts the interpretation that the description refers to ensembles of systems and not to individual systems. In that case the whole "egg-walking" performed in order to avoid the "physically real" becomes superfluous. There exists, however, a simple psychological reason for the fact that this most nearly obvious interpretation is being shunned. For if the statistical quantum theory does not pretend to describe the individual system (and its development in time) completely, it appears unavoidable to look elsewhere for a complete description of the individual system; in doing so it would be clear from the very beginning that the elements of such a description are not contained within the conceptual scheme of the statistical quantum theory. With this one would admit that, in principle, this scheme could not serve as the basis of theoretical physics. Assuming the success of efforts to accomplish a complete physical description, the statistical quantum theory would, within the framework of future physics take an approximately analogous position to the statistical mechanics within the framework of classical mechanics. I am rather firmly convinced that the development of theoretical physics will be of this type; but the path will be lengthy and difficult.

I now imagine a quantum theoretician who may even admit that the quantum-theoretical description refers to ensembles of systems and not to individual systems, but who, nevertheless, clings to the idea that the type of description of the statistical quantum theory will, in its essential features, be retained in the future. He may argue as follows: True, I admit that the quantum-theoretical description is an incomplete description of the individual system. I even admit that a complete theoretical description is, in principle, thinkable. But I consider it proven that the search for such a complete description would be aimless. For the lawfulness of nature is thus constituted that the laws can be completely and suitably formulated within the framework of our incomplete description.

To this I can only reply as follows: Your point of view—taken as theoretical possibility—is incontestable. For me, however, the expectation that the adequate formulation of the universal laws involves the use of *all* conceptual elements which are necessary for a complete description, is more natural. It is furthermore not at all surprising that, by using an incomplete description, (in the main) only statistical statements can be obtained out of such description. If it should be possible to move forward to a complete description, it is likely that the laws would represent relations among all the conceptual elements of this description which, per se, have nothing to do with statistics.

Bohr on Complementarity

Neils Bohr's principle of complementarity is an explanatory goal that stipulates that a *complete* explanation of a quantum mechanical process include "descriptions" of what happens to a system between observations. On the Copenhagen interpretation developed by Bohr and Heisenberg there are three levels of interpretation of an experiment in the quantum realm. Level one is the description of the experimental arrangement and results. Level two is a set of calculations made using the formalism of quantum theory—either Heisenberg's matrix mechanics or Schrödinger's wave equations. It is at level two that predictions are made about the re- sults of further observations to be made of the system. Level three con- tains mutually exclusive, but complementary "descriptions" of quantum mechanical systems during the intervals between measurements.

Level three "descriptions" do not enter into the calculations of elec- tron transition frequencies, scattering distributions, or other quantum phenomena. Nevertheless, Bohr and Heisenberg insisted that a *complete* understanding of quantum phenomena include them. The principle of complementarity specifies that two mutually exclusive descriptions of a quantum process, in which two different experimental arrangements are employed, are complementary and combine to produce an exhaustive in- terpretation of the process. In the case of the diffraction of electrons by a single slit, the two mutually exclusive descriptions are as follows: (1) "elec- trons which have a certain position in passing through a diaphragm pro- duce a certain pattern of impacts on a photographic plate," and (2) "elec- trons which have a certain momentum in passing through a diaphragm transfer a certain momentum to the diaphragm." On the Copenhagen in- terpretation, these two descriptions combine to exhaust the range of pos- sible knowledge about the passage of electrons through a single slit in a diaphragm.

Level three interpretations may be given in terms of the classical con- cepts of "waves" and "particles." In the single-slit experiment, description

(1) refers to "electrons *qua* waves," whereas description (2) refers to "electrons *qua* particles." The two pictures are mutually exclusive. A particle is localizable in space and time. A pure wave is not, but is infinitely extended in space and time. It is for this reason that Max Born declared that "the imagination can scarcely conceive two ideas which appear less capable of being united than these two, which the quantum theory proposes to bring into such close connection."[10]

Level three interpretations also may be given in terms of (1) the "paths" followed between observations, and (2) interactions in which energy and momentum are conserved. These two types of interpretation also are mutually exclusive. As Bohr pointed out "we are presented with a choice of *either* tracing the path of a particle *or* observing interference effects."[11]

Within classical physics the results of a collision process, for instance the collision of billiard balls, may be interpreted *both* by specifying the paths taken in space and time *and* by providing a causal analysis of the interaction. The causal analysis invokes the principles of conservation of energy and conservation of momentum.

Bohr maintained that spatiotemporal description and causal analysis are mutually exclusive interpretations within quantum physics. He declared that "the impossibility of a separate control of the interaction between the atomic objects and the instruments indispensable for the definition of the experimental conditions prevents the unrestricted combination of space-time coordination and dynamical conservation laws on which the deterministic description in classical physics rests."[12]

The principle of complementarity stipulates that pictorial-level interpretations of quantum phenomena in wave language and in particle language are of equivalent importance. This equivalence is based ultimately on the uncertainty relations, Einstein's equation $E = hv$, and De Broglie's equation $p = hv/c$. If the equations expressing energy and momentum in terms of frequency are substituted into the uncertainty relations $\Delta p \, \Delta q \geq h/4$ and $\Delta E \, \Delta t \geq h/4$, it is evident that there is a formal symmetry between the specification of spatiotemporal coordinates and the specification of wavelike properties. Bohr and Heisenberg emphasized that these symmetry properties are the foundation of quantum mechanics.

Complementarity is an essential doctrine within the Copenhagen interpretation of quantum mechanics. Bohr defended this principle on the grounds that the extension of physical theory to new areas of experience ought to be grounded in the language used to describe everyday experience. He insisted that the objective description of experience requires unambiguous communication, and that the use of "plain language," which serves the needs of everyday living, is a necessary condition of unambiguous communication. Bohr declared that "the development of atomic physics has taught us how, without leaving common language, it is possible to create a framework sufficiently wide for an exhaustive description of experience.[13]

Critics of the Copenhagen interpretation were not impressed by this defense. Einstein, Schrödinger, Margenau, Bohm, and numerous others have rejected complementarity as an explanatory ideal. Critics and supporters of the Copenhagen interpretation agree on how to apply the formalism of quantum theory to predict the results of experiments. The critics, however, seek to formulate interpretations of the formalism that take either waves (e.g., Schrödinger) or particles (e.g., Bohm) to be fundamental, thereby rejecting the Copenhagen ideal of irreducible wave-particle dualism.

The "Ionian Enchantment"

Scientists not only express preferences for specific types of explanation, they also defend general explanatory goals. Holton called attention to one such goal, a goal he labeled the "Ionian Enchantment."[14] The Ionian Enchantment is the ideal of a unified theory of all phenomena. This has been an important goal of theoretical physics from the universal force curve of Roger Boscovich (1763)[15] to the present search for a "theory of everything" that will integrate and unify the four fundamental forces—strong nuclear, weak nuclear, electromagnetic, and gravitational. To subscribe to the Ionian Enchantment is to favor scientific explanations that invoke the fewest and most inclusive premises. Einstein expressed this goal as follows:

The aim of science is, on the one hand, a comprehension, as *complete* as possible, of the connection between the sense experiences in their totality, and, on the other hand, the accomplishment of this aim *by the use of a minimum of primary concepts and relations*. (Seeking, as far as possible, logical unity in the world picture, i.e. paucity in logical elements.)

*   *   *

The essential thing is the aim to represent the multitude of concepts and theorems, close to experience, as theorems, logically deduced and belonging to a basis, as narrow as possible, of fundamental concepts and fundamental relations which themselves can be chosen freely (axioms). The liberty of choice, however, is of a special kind; it is not in any way similar to the liberty of a writer of fiction. Rather it is similar to that of a man engaged in solving a well designed word puzzle. He may, it is true, propose any word as the solution; but, there is only *one* word which really solves the puzzle in all its forms. It is an outcome of faith that nature—as she is perceptible to our five senses—takes the character of such a well formulated puzzle. The successes reaped up to now by science do, it is true, give a certain encouragement for this faith.[16]

Ontological Commitments and High-Level Hypotheses

Holton also included among themata ontological commitments (e.g., to atoms, caloric, aether, and so on) and commitments to high-level hypotheses. High-level hypotheses include the constancy of the velocity of light, the discreteness of electric charge, and the quantization of energy. Acceptance of these hypotheses influences evaluative practice in science.

THEMATIC PRINCIPLES AND THE HISTORY OF SCIENCE

Thematic principles, however widely held, are subject to modification and even abandonment. Commitments to the conservation of mass and parity, and the existence of phlogiston, caloric, and aether once were influential in theory evaluation. However, these thematic principles no longer are accepted within the scientific community.

Despite their susceptibility to change, thematic principles are very much involved in theory appraisal. Holton suggested, moreover, that reference to the thematic dimension resolves several puzzles about the historical development of science.

It has been well-documented that science is a cooperative and cumulative enterprise, that eminent scientists often disagree about the importance of particular research programs, and that successful theorists often "suspend disbelief" in the face of negative evidence. Holton maintained that these aspects of the scientific enterprise are best understood by reference to thematic considerations.[17]

In the first place, scientists' shared commitments to certain thematic principles make science a cooperative, cumulative undertaking. Scientists often agree about the types of theory to be sought and the types of explanation that are relevant. This agreement promotes continuity within research.

It also is true, however, that commitments to thematic principles sometimes promote conflict. Certain thematic principles pull in opposite directions. For example, atomism and plenism are mutually exclusive ontological positions, and the evaluative standards simplicity and agreement with observations are antithetical. Commitments to opposed thematic principles underlie the disputes between such gifted scientists as Newton and Leibniz, Bohr and Einstein, and Heisenberg and Schrödinger.

Reference to thematic principles also casts light upon the propensity of scientists to perform a "suspension of disbelief." Holton called attention to Einstein's response to Walter Kaufmann's experimental result (1906) that appeared to falsify the special theory of relativity. Einstein refused to accept the finding as decisive. According to Holton, he did so largely because he was strongly committed to the thematic principles of simplicity, unification, symmetry, and invariance.[18] Had Einstein not been convinced that fulfilling these formal requirements is of overriding importance, he presumably would have attributed greater weight to the *prima facie* destructive implications of Kaufmann's work.

IS IT TRUE THAT "ANYTHING GOES"?

It might seem from the foregoing that there are no immutable standards that govern the evaluation of theories. This was the conclusion of

Paul Feyerabend. Feyerabend promoted a methodological anarchism according to which "anything goes."[19]

FEYERABEND ON SCIENCE AND RULES

The idea of a method that contains firm, unchanging, and absolutely binding principles for conducting the business of science meets considerable difficulty when confronted with the results of historical research. We find, then, that there is not a single rule, however plausible, and however firmly grounded in epistemology, that is not violated at some time or other. It becomes evident that such violations are not accidental events; they are not results of insufficient knowledge or of inattention which might have been avoided. On the contrary, we see that they are necessary for progress. Indeed, one of the most striking features of recent discussions in the history and philosophy of science is the realization that events and developments, such as the invention of atomism in antiquity, the Copernican Revolution, the rise of modern atomism (kinetic theory; dispersion theory; stereochemistry; quantum theory), the gradual emergence of the wave theory of light, occurred only because some thinkers either *decided* not to be bound by certain "obvious" methodological rules, or because they *unwittingly broke* them.

This liberal practice, I repeat, is not just a *fact* of the history of science. It is both reasonable and *absolutely necessary* for the growth of knowledge. More specifically, one can show the following: given any rule, however "fundamental" or "necessary" for science, there are always circumstances when it is advisable not only to ignore the rule, but to adopt its opposite. For example, there are circumstances when it is advisable to introduce, elaborate, and defend *ad hoc* hypotheses, or hypotheses which contradict well-established and generally accepted experimental results, or hypotheses whose content is smaller than the content of the existing and empirically adequate alternative, or self-inconsistent hypotheses, and so on.

I think Feyerabend is guilty of overstatement. The game of science has been played according to rules. Some of these rules do have the status of inviolable principles.

Intratheoretical consistency is the most important inviolable principle applicable to theory choice. No acceptable scientific theory can violate this principle. Scientists and philosophers debate the value of evaluative standards such as undesigned scope and complementarity. They do not debate the inviolable status of intratheoretical consistency. A person who rejects this standard would be regarded as having opted out of the game of science.

It is important to distinguish intratheoretical consistency from intertheoretical consistency. Scientists do not require that a new theory be consistent with established theories in order to be acceptable. But they invariably require, and should require, that a theory be internally consistent. Intratheoretical consistency is a necessary condition of cognitive significance. If a theory includes inconsistent postulates, then it implies any statement whatever. But a theory that implies both *S* and *not S* provides support for neither. I suppose one may conceive of epistemically desperate situations in which consistency is no more advantageous than inconsistency. Under such conditions life would be precarious indeed. Confronted with a need to take action, an individual could receive no guidance from scientific theories.

One might seek to justify the inviolable status of intratheoretical consistency by appeal to the theory of organic evolution. Intratheoretical consistency is justified because its applications have proved to be adaptive for *Homo sapiens.*

This appeal to evolution is not decisive, however. Three types of objections may be raised. The first objection, to parody Feyerabend, is "what is so great about *Homo sapiens?*" Granted that a type of response to environmental pressures has proved adaptive for *Homo sapiens,* why should we take the continued success of *Homo sapiens* to be the measure of evolutionary development?

The second objection is that the evolutionary defense involves an element of circularity. Commitment to intratheoretical consistency is justified by appeal to evolutionary theory, which theory itself is justified, in part, by appeal to intratheoretical consistency.

The third objection is perhaps the most damaging. "Evolutionary success" requires both adaptation and the retention of adaptability. It always is possible that adaptation in a specific context has been achieved in such a way as to diminish the capacity to respond creatively to further changes of environmental conditions. Adaptation may have been purchased at the expense of a loss of adaptability.

It is possible, at least, that this is the situation with respect to a methodological standard hitherto accepted as foundational. The evolutionist cannot prove that adaptability has been retained. That adaptability has been retained can be established only by reference to what happens at subsequent times. Since the above objections have some force, the evolutionary argument to justify intratheoretical consistency is not effective.

John Worrall, citing Popper and Lakatos, has insisted that "ultimately we must stop arguing and 'dogmatically' assert certain basic principles of rationality."[20] The process of justification by appeal to more basic principles must stop at some point. Affirmation of intratheoretical consistency as a condition of scientific rationality is such a point.

A second inviolable principle applicable to scientific theories is that every acceptable theory must include among its terms at least one term linked, however indirectly, to instrumental procedures that determine its values. This operational requirement stipulates that an acceptable theory have empirical consequences.

The overwhelming majority of theories proposed by scientists satisfy this requirement. An exception is Newton's theory of color mixing.[21] Newton suggested that the resultant color produced by mixing spectral pigments may be calculated from ratios among intervals in the musical octave. He assigned each of the seven spectral colors to a wedge within a circle. Newton took the widths of the seven wedges to be proportional to the musical intervals in the octave. He stipulated that the "number of rays" of each component color be represented by a circle of the appropriate size located at the midpoint of the arc of that pigment's segment. Newton

Figure 24. Newton's Theory of Color Mixing.
Isaac Newton, *Opticks* (New York: Dover, 1952),
p. 155.

maintained that the "center of gravity" of the small circles falls within the color segment of the resulting mixture.

Newton failed to specify how to measure the "number of rays" of a pigment. Consequently, the assignment of radii to the small circles is arbitrary. Newton's theory fails the operational requirement.

It must be granted that intratheoretical consistency and the operational requirement place only minimal constraints on the evaluative process. To insist that acceptable theories be internally consistent and have empirical consequences disqualifies very few candidates. What scientists wish to know is whether empirically significant, consistent theory T_1 is superior to empirically significant, consistent theory T_2.

Worrall maintained that the following evaluative principles also have inviolable status within science:

1. Theories should be tested against plausible rivals (if there are any).[22]

2. Non–*ad hoc* theories should always be preferred to *ad hoc* ones (where both are available).[23]

3. Greater empirical support can legitimately be claimed for the hypothesis that a particular factor caused some effect if the experiment testing the hypothesis has been shielded against other possible causal factors.[24]

Worrall maintained that these principles underlie the practice of science in the same way that *modus ponens* underlies deductive logic. A person who accepts p and $p \supset q$ but refuses to accept q (e.g., Lewis Carroll's tortoise)[25] has opted out of the game of deductive logic. According to Worrall, a person who repudiates these inviolable principles likewise has opted out of the game of science.

Worrall may be correct that the game of science is played according to these rules. Unfortunately, appeal to these rules decides the issue in relatively few cases of theory choice. In most instances scientists apply more specific evaluative standards—simplicity, accuracy, predictive success, inclusiveness, consilience, undesigned scope, and so on. But as Kuhn emphasized, these standards conflict and there is disagreement about their relative importance.

Kuhn is correct that the history of science reveals no agreed upon algorithm for theory choice. There is considerable latitude for rational differences of opinion within evaluative contexts. Nevertheless, it always is relevant to ask questions that invoke the above standards, e.g.,

1. Do calculations made using T agree sufficiently well with predictions or past observations?
2. Was the extension of T to account for p unexpected?
3. Does T explain everything explained by T_{n-1}?
4. Is T too complex?
5. Should T, which does not account for p, have been expected to account for p?

Scientists regard negative answers to questions 1, 2, and 3, and affirmative answers to questions 4 and 5, to count against the acceptability of T. Of course, other considerations may be relevant as well. But scientific evaluative practice is not anarchic. It is not the case that "anything goes."

Conclusion

It is appropriate at this point to review the conclusions reached about the questions posed in the introduction. Consider, first of all, question 2: Are theories ever falsified directly upon appeal to empirical evidence? There is general agreement that Duhem is correct about the logic of this relationship. What is falsified by an empirical result is not a hypothesis alone, but rather the conjunction of the hypothesis and a statement about relevant conditions. One may salvage a hypothesis by taking the statement about relevant conditions to be false.

However, in practice, scientists do take negative experimental evidence to falsify low-level hypotheses. A typical instance is the identification of an organic compound. Many such compounds have distinctive, sharp melting points. The failure of an organic substance (e.g., sucrose) to melt at a particular temperature (186°C) is taken to falsify the hypothesis "that substance is sucrose." Of course, it is possible that the investigator has determined the melting point incorrectly. But, in practice, scientists routinely assume that such determinations are correct. And as Popper has emphasized, the acceptance of some "basic statements" as true is necessary in order for there to be genuine tests of hypotheses. According to Popper, it is the continuing exposure to testing that distinguishes science from other human endeavors.

The case of high-level theories is more complex. Potentially falsifying evidence is related to such theories via a network of statements about initial conditions and boundary conditions, auxiliary hypotheses and (usually implicit) assumptions about the operation of scientific instruments. Quine was correct to emphasize that scientists may respond to negative evidence in a variety of ways. Usually they seek to achieve accommodation to the evidence by modifying a "peripheral" aspect of the theoretical network. But occasionally scientists opt for fundamental changes that drastically alter the basic postulates of a theory. For example, Leverrier and Adams responded to evidence of the anomalous motion of Uranus by modifying the set of relevant conditions that affect the planet's motion. They postulated the existence of a new planet. G. B. Airy, by contrast, was prepared to alter the law of gravitational attraction, a basic postulate of Newtonian planetary theory.[1]

The claim that theory T is made false by evidence statement e cannot be sustained. Nevertheless, theories are rejected, and some rejections contribute to scientific progress. Question 3 raises the issue of what conditions are necessary or sufficient for theory rejection.

Francis Bacon directed attention to "crucial experiments" that supposedly decide the issue between two competing theories. Many eighteenth-century and early nineteenth-century methodologists echoed Bacon's emphasis. In the 1850s, there was considerable support for the claim that the Fizeau-Foucault experiment, which established that the velocity of light is greater in air than in water, demonstrated that the wave theory is true and the corpuscular theory is false. To take this position is to presuppose that the wave and corpuscular theories are the only two possibilities. This presupposition was shown to be false upon formulation of the electromagnetic theory of light.

Duhem was correct to conclude that, since one cannot prove that there are only two viable competing theories for some class of phenomena, there are no crucial experiments. But even if no experiment satisfies the philosopher's requirements for "crucial" status, scientists are impressed by those relatively crucial experiments that provide support for one theory but not for its competitor. On many occasions they take such results to be a sufficient reason to reject the lesser competing theory.

Thomas Kuhn maintained that a scientific theory is rejected only after a viable competitor arrives on the scene. This view about theory rejection must be rejected itself. Some high-level theories in the history of science have been rejected in the absence of a recognized competitor (chapter 8). Competition with a second theory is not a necessary condition of theory rejection. Of course, this is not to deny that competition often does precede theory rejection.

There remains question 5—"What determines whether theory replacement is justified?" Whewell, Lakatos, and others have provided answers to this question. They shared the philosopher's typical concern to specify necessary and/or sufficient conditions.

Whewell recommended "undesigned scope" as a sufficient condition of justified theory replacement. A display of undesigned scope reveals that a theory is applicable to a new class of phenomena. Methodologists have expressed doubts about our ability to determine if effects are "different in kind." And even when there is agreement that a theory has been extended to effects "different in kind," methodologists are divided about its importance (chapter 11). In any case, claims about the achievement of undesigned scope have been made for relatively few historical episodes.

"Consilience" (Whewell) and "incorporation with corroborated excess content" (Lakatos) fail as sufficient conditions as well. There are countercases—episodes within the history of science which display consilience or incorporation with corroborated excess content without contributing to scientific progress. To those preoccupied with formal requirements this is unsettling. However, it is perfectly reasonable to acknowledge the importance of these criteria—to highlight the many episodes that conform to these requirements—while acknowledging that satisfaction of one of these criteria is not a guarantee of progressive theory replacement.

Kuhn has outlined the difficulties that confront those who would impose formal criteria upon the task of theory appraisal. Applications of the principal recommended criteria—agreement with observations, simplicity, breadth of scope, fertility—conflict (see chapter 8). Each criterion justifies the transition $T_1 \Rightarrow T_2$ only if no other considerations favor T_1. But this situation is seldom realized. The demands of descriptive-predictive accuracy and explanatory power usually pull in opposite directions. Kuhn and Holton have insisted, moreover, that any reconstruction of the evaluative practice of scientists take account of "idiosyncratic factors" (Kuhn) and "themata" (Holton).

The methodologist committed to the specification of conditions necessary or sufficient for justified theory replacement is not without recourse to inviolable standards, however. Intratheoretical consistency and links to operationally defined terms are conditions necessary for acceptable scientific theories. And Worrall is correct to include among inviolable criteria the instructions to prefer non–*ad hoc* theories, prefer theories that have been tested against plausible rivals, and prefer theories about causal connections whose tests have been shielded from other possible causal factors. Unfortunately, these formal considerations place only modest restrictions on acceptable theories.

Notes

Introduction

1. Charles Darwin, Autobiography, in *The Life and Letters of Charles Darwin,* ed. Francis Darwin (New York: Basic Books, 1959) vol. 1, 68.

2 See, for instance, Michael T. Ghiselin, *The Triumph of the Darwinian Method* (Berkeley: University of California Press, 1969); Michael Ruse, *The Darwinian Revolution* (Chicago: University of Chicago Press, 1979), chapter 7.

Chapter 1. The Logic of Falsification

1. Aristotle, *Generation of Animals,* III, V, 756b.

2. Ibid., IV, i, 764a–764b.

3. Stephen Hales, *Vegetable Staticks,* 1727 (London, Oldbourne, 1961), 73, 128.

4. Count Rumford, "Source of the Heat which is Excited by Friction" (1798), in *The Collected Works of Count Rumford, vol. 1* (Cambridge, Mass.: Harvard University Press, 1968), 4–6, 14–15, 20–22.

5. Pierre de Maupertuis, *Lettres de M. de Maupertuis* (Dresden, 1751), vol. 2, letter 14. Quoted in Elisabeth Gasking, *Investigations Into Generation, 1651–1828* (London: Hutchinson, 1967), 79, 81.

6. See, for instance, Elizabeth Gasking, *Investigations into Generation, 1651–1828,* ch. 2–4.

7. Maupertuis, *Lettres,* 79.

8. These developments are the subject of a classic study of the chemical revolution by James B. Conant, *The Overthrow of the Phlogiston Theory: Harvard Case Studies in Experimental Science, Case 2* (Cambridge, Mass.: Harvard University Press, 1956).

9. Joseph Priestley, "Experiments and Observations on Different Kinds of Air," in *The Discovery of Oxygen, Part 1* (Edinburgh: Alembic Club Reprint No. 7, 1961), 16–19.

10. Ernest Rutherford, "Forty Years of Physics," in *Background to Modern Science,* ed. J. Needham and W. Pagel (New York: MacMillan, 1938), 68–69.

11. Isaac Newton, *Mathematical Principles of Natural Philosophy,* vol. 1, trans. F. Cajori (Berkeley: University of California Press, 1962), 393–95.

12. See, for instance, E. J. Aiton, *The Vortex Theory of Planetary Motion* (New York: American Elsevier, 1972).

Chapter 2. The Limits of Falsification

1. Pierre Duhem, *The Aim and Structure of Physical Theory, 1914* (New York: Atheneum, 1962), 183–87.

2. Ibid.

3. Willard van Orman Quine, "Two Dogmas of Empiricism," in *From a Logical Point of View* (Cambridge, Mass.: Harvard University Press, 1953), 42–45.

4. Theodocius Dobzhansky, *Evolution, Genetics, and Man* (New York: Wiley, 1963), 360.

5. Stephen J. Gould, "The Origin and Function of 'Bizarre' Structures: Antler Size and Skull Size in the 'Irish Elk', *Megaloceros Giganteus*," *Evolution 28* (1974), 216.

6. George Gaylord Simpson, *The Meaning of Evolution* (New Haven: Yale University Press, 1949), 150.

7. Gould, "The Origin and Function of 'Bizarre' Structures," 192–213.

Chapter 3. Are There Crucial Experiments?

1. Francis Bacon, *Novum Organum*, book 2, aphorism 36, no.5.

2. Richard Leakey, *The Origin of Humankind* (New York: Basic Books, 1994), 87, 95–97.

3. Isaac Newton, *Philosophical Transactions 80* (1671/1672), 3078; reprinted in *Isaac Newton's Papers and Letters on Natural Philosophy*, ed. I. B. Cohen (Cambridge, Mass.: Harvard University Press, 1958), 50.

4. Newton, *Opticks*, 4th ed., 1730 (New York: Dover, 1952), 26–48.

5. Robert Hooke, "Letter to Lord Brouncker," June 1672, in *The Correspondence of Isaac Newton*, vol. 1, ed. H. W. Turnbull (Cambridge: Cambridge University Press, 1959), 202.

6. Newton, *Opticks*, 26.

7. Duhem, *The Aim and Structure of Physical Theory, 1914* (New York: Atheneum, 1962), 188–90.

8. John Herschel, *A Preliminary Discourse on the Study of Natural Philosophy*, 1830 (New York: Johnson Reprint, 1966), 229–300.

9. Florin Perier, "Letter Sent by M. Perier to M. Pascal the Younger," September 22, 1648, in *The Physical Treatises of Pascal*, trans. I. H. B. and A. G. H. Spiers (New York: Octagon Books, 1973), 104–5.

10. Robert Brandon, *Adaptation and Environment* (Princeton, N.J.: Princeton University Press, 1995), 178–79.

11. Joel Kingsolver and M. A. R. Koehl, "Aerodynamics, Thermoregulation, and the Evolution of Insect Wings: Differential Scaling and Evolutionary Change," *Evolution 39* (1985), 489–90.

12. Ibid., 501.

13. Ibid., 489–91; 500–503. Citations omitted in quoted passage.

Chapter 4. Falsification and the Method of Difference

1. John Stuart Mill, *A System of Logic*, 8th ed. (London: Longman, 1970), book 3, chapter 8, 256.

2. Albert Einstein, *Relativity* (New York: Crown, 1961), 52–54.

3. For a comprehensive discussion of the Michelson-Morley results and the re-

sponses of G. F. Fitzgerald, Henrik Lorentz, and Albert Einstein, see Gerald Holton, *Thematic Origins of Scientific Thought: Kepler to Einstein* (Cambridge, Mass: Harvard University Press, 1973), chapter 9.

4. Mill, *A System of Logic,* book 3, chapter 9, 267–68.

5. Ibid., chapter 11, 299–305.

6. Ibid., chapter 14, 323–24.

7. Newton had not been able to demonstrate this stability. He appealed to the continuing activity of God as governor of the universe to overcome perturbation effects and maintain the integrity of planetary orbits. Isaac Newton, *Mathematical Principles of Natural Philosophy,* trans. F. Cajori (Berkeley: University of California Press, 1962), vol. 2, 544.

Chapter 5. Popper's Methodological Falsificationism

1. Karl Popper, *The Logic of Scientific Discovery* (New York: Basic Books, 1959), 108–9.

Chapter 6. Responses to *Prima Facie* Falsifying Evidence

1. Dmitri Mendeleyev, "The Periodic Law of the Chemical Elements," *Chemical News,* 40, 1045 (Dec. 5, 1879), 267. Reprinted in *Classical Scientific Papers—Chemistry, Second Series,* ed. D. M. Knight (New York: Elsevier, 1970), 279.

2. Mendeleyev, "The Periodic Law of the Chemical Elements," *Chemical News* 41, 1057 (Feb. 27, 1880), 93–94. Reprinted in *Classical Scientific Papers—Chemistry, Second Series,* 301–2.

3. See J. W. van Spronsen, *The Periodic System of the Chemical Elements* (New York: Elsevier, 1969) 137.

4. Ibid., 139.

5. Galileo, *Dialogues Concerning Two New Sciences,* trans. H. Crew and A. de Salvio (New York: Dover, 1914), 254–55; 85. *Dialogue Concerning the Two Chief World Systems,* trans. S. Drake (Berkeley: University of California Press, 1953), 450.

6. Double refraction was described by Erasmus Bartholin in 1669. See William Whewell, *History of the Inductive Sciences,* 1857 (London: Cass, 1967), part 2, vol. 3, 293–94.

7. Antony Flew, "Theology and Falsification," in *New Essays in Philosophical Theology,* ed. A. Flew and A. MacIntyre (London: SCM Press, 1955), 96–99; John Wisdom, "Gods," in *Logic and Language: First Series,* ed. A. Flew (Garden City: Doubleday Anchor, 1965), 200–202.

8. Flew, "Theology and Falsification," 97.

9. Popper, *The Logic of Scientific Discovery* (New York: Basic Books, 1959), 82–83.

10. J. B. A. Dumas, "Memoir on the Equivalents of the Elements," *Ann. Chem. Phys.* 55 (1859), 129–210.

11. Popper, *Objective Knowledge* (Oxford: Clarendon, 1972), 14–15, 83.

Chapter 7. Whewell on Scientific Progress

1. William Whewell, *History of the Inductive Sciences,* 3 vols, 3rd ed., 1857 (London: Cass, 1967).

2. Whewell, *Philosophy of the Inductive Sciences,* 2nd ed., 1847 (London: Cass, 1967), vol. 2, 85.

3. A good example of the "explication" of a conception is the definition of the instantaneous velocity of a body:

$$v_i = \lim_{t \to o} \frac{s}{t}$$

Fourteenth-century theorists at Merton College, Oxford made significant contributions to the science of dynamics by reference to this newly clarified concept. See, for instance, Marshall Clagett, *The Science of Mechanics in the Middle Ages* (Madison: University of Wisconsin Press, 1959), 199–329. The progressive explications of such concepts as "force," "temperature," and "gene" also have contributed to the growth of science.

4. Whewell, *Philosophy of the Inductive Sciences,* vol. 2, book 11, 118.

5. Ibid., 65.

6. John Herschel, *A Preliminary Discourse on the Study of Natural Phenomena,* 1st ed., 1830 (New York: Johnson, 1966), 33–34, 171–72.

7. Whewell, *Philosophy of the Inductive Sciences,* vol. 2, 67–68.

8. Herschel, *A Preliminary Discourse on the Study of Natural Philosophy,* 171–72; Whewell, *Philosophy of the Inductive Sciences,* vol. 2, book 11, 67–68.

Chapter 8. Kuhn on Theory Replacement

1. Paul Dirac, *The Principles of Quantum Mechanics* (Oxford: Clarendon Press, 1930). Reprinted in *The World of the Atom,* ed. H. A. Boorse and L. Motz (New York: Basic Books, 1966), 1175–78.

2. Thomas S. Kuhn, *The Structure of Scientific Revolutions,* 2nd ed. (Chicago: University of Chicago Press, 1970), 77.

3. Ibid.

4. Imre Lakatos, "Falsification and the Methodology of Scientific Research Programmes," in *Criticism and the Growth of Knowledge,* ed. I. Lakatos and A. Musgrave (Cambridge: Cambridge University Press, 1970), 119.

5. C. N. Yang, *Elementary Particles* (Princeton, N.J.: Princeton University Press, 1962), 53–58.

6. See, for instance, Robert Fox, "The Rise and Fall of Laplacian Physics," *Hist. Stud. Phys. Sci.* 4 (1974), 89–136.

7. H. C. Oersted, "Experiments on the Effect of a Current of Electricity on the Magnetic Needle" (1820), in *Source Book in Physics,* ed. W. F. Magee (Cambridge, Mass.: Harvard University Press, 1963), 438–39.

8. See, for instance, Robert Fox, "The Rise and Fall of Laplacian Physics."

9. Kuhn, *The Structure of Scientific Revolutions,* 150.

10. Jack Meiland, "Kuhn, Scheffler, and Objectivity in Science," *Phil. Sci.* 41 (1974), 183.

11. Kuhn, *The Essential Tension* (Chicago: University of Chicago Press, 1977), 322.

12. Ibid., 331.

13. Ibid., 329.

Chapter 9. Lakatos on Progressive Research Programs

1. Imre Lakatos, "Falsification and the Methodology of Scientific Research Programmes," in *Criticism and the Growth of Knowledge*, ed. I. Lakatos and A. Musgrave (Cambridge: Cambridge University Press, 1970), 116–18.

2. D. Mendeleyev, "The Periodic Law of the Chemical Elements," *Chemical News* 40, no. 1045 (Dec. 5, 1879), 268. The atomic weights assigned to the elements are recent values.

3. Lakatos, "History of Science and Its Rational Reconstructions," in *Boston Studies in the Philosophy of Science*, vol. 8, ed. R. Buck and R. S. Cohen (Dordrecht: Reidel, 1971), 104–5.

4. Ibid., 100–101.

5. Peter Clark, "Atomism *versus* Thermodynamics," in *Method and Appraisal in the Physical Sciences*, ed. Colin Howson (Cambridge: Cambridge University Press, 1976), 45–46; 47–49; 59–60; 60–61.

6. Ibid.

7. Alan Musgrave, "Why Did Oxygen Supplant Phlogiston? Research Programmes in the Chemical Revolution," in *Method and Appraisal in the Physical Sciences*, 187–91; 203.

8. Isaac Newton, *Opticks*, 1730 (New York: Dover, 1952), 280–82.

9. Lakatos, "Falsification and the Methodology of Scientific Research Programmes," 138–40.

10. Dudley Shapere, "The Character of Scientific Change," in *Scientific Discovery, Logic and Rationality*, ed. T. Nickles (Dordrecht: Reidel, 1980), 83–84.

11. For instance, logicians have developed competing formal systems for calculating values of the "degree of confirmation" of hypotheses. What has yet to be shown is that some one of these systems is appropriate for a wide range of scientific practice.

12. Lakatos, "History of Science and Its Rational Reconstructions," 108–22.

13. Musgrave, "Method or Madness?" in *Boston Studies in the Philosophy of Science*, vol. 39, ed. R. S. Cohen, et al. (Dordrecht: Reidel, 1976), 458; 459–62; 466–67.

14. Lakatos, "History of Science and Its Rational Reconstructions," 105.

15. Thomas S. Kuhn, "Notes on Lakatos," in *Boston Studies in the Philosophy of Science*, vol. 8, ed. R. Buck and R. S. Cohen (Dordrecht: Reidel, 1971), 141–46.

Chapter 10. The Asymptotic Agreement of Calculations

1. R. E. Peierls, *The Laws of Nature* (New York: Scribner's, 1956), 121–22.

2. Copernicus explained the absence of stellar parallax by invoking the unsupported assumption that the angle of parallax is too small to be measured. Subsequently, Galileo argued that the fall of bodies to the base of towers does not count against the rotation of the earth because the observed motion of a body is not its "real motion."

3. Niels Bohr, *Atomic Theory and the Description of Nature* (Cambridge: Cambridge University Press, 1961), 36–37.

4. Ernest H. Hutten, *The Language of Modern Physics* (London: George Allen & Unwin, 1956), 165–66; 168.

5. A black body is an idealized isothermal radiator that absorbs 100 percent of

the energy incident upon it. A good approximation to this ideal is a box lined with lampblack with a small hole in the center of one wall.

6. See, for instance, F. W. Van Name, *Modern Physics* (New York: Prentice Hall, 1952), 45–49.

7. Max Jammer, *The Conceptual Development of Quantum Mechanics* (New York: McGraw-Hill, 1966), 23.

Chapter 11. Successful Prediction and the Acceptability of Theories

1. R. E. Peierls, *The Laws of Nature* (New York: Charles Scribner's Sons, 1956), 121.

2. George Gamow, *One, Two, Three . . . Infinity* (New York: Viking Press, 1948), 302–3.

3. Hans Bethe, "Energy Production in Stars," *Phys. Rev.* 55 (1939). Reprinted in *The World of Atoms,* ed. H. A. Boorse and L. Motz (New York: Basic Books, 1966), 1634–36.

4. Imre Lakatos, "Changes in the Problem of Inductive Logic," in *Inductive Logic,* ed. I. Lakatos (Amsterdam: North-Holland, 1968), 376–77.

5. See Vasco Ronchi, *The Nature of Light* (London: Heinemann, 1970), 221.

6. James Clerk Maxwell, "Illustrations of the Dynamical Theory of Gases" *Phil. Mag.* 1859, 19–20. Reprinted in *Kinetic Theory,* ed. S. Brush (Oxford: Pergamon Press, 1948), 148–49.

7. Fred J. Vine, "Spreading of the Ocean Floor: New Evidence," *Science* 154 (1966), 1405–6.

8. William Whewell, *History of the Inductive Sciences,* 3rd ed., 1857 (London: Cass, 1967), vol. 2, 250.

9. Albert Einstein and Leopold Infeld, *The Evolution of Physics* (New York: Simon and Schuster, 1961), 258–62.

10. Martin Carrier, "What Is Wrong with the Miracle Argument?" *Stud. Hist. Phil. Sci.* 22 (1991), 30.

11. Ibid.

12. Ibid., 31.

13. John Worrall, "Fresnel, Poisson and the White Spot: The Role of Successful Predictions in the Acceptance of Scientific Theories," in *The Uses of Experiment,* ed. D. Gooding, T. Pinch, and S. Schaffer (Cambridge: Cambridge University Press, 1989), 142–46.

14. Stephen G. Brush, "Dynamics of Theory Change: The Role of Predictions," *PSA: Proceedings of the Biennial Meeting of the Philosophy of Science Association,* vol. 2 (1994), 138.

15. Peter Achinstein, "Explanation v. Prediction: Which Carries More Weight?" *PSA,* vol. 2 (1994), 163.

Chapter 12. Variety within Evidence

1. Stanislao Cannizzaro, *Sketch of a Course of Chemical Philosophy,* 1858 (Edinburgh: Alembic Club Reprint no. 18, 1969), 11–13.

2. Jean Perrin, *Atoms,* 1913, trans. D. Hammick (London: Constable, 1922), 215–17.

3. Max Planck, "The Origin and Development of the Quantum Theory," Nobel Prize in Physics Award Address, 1920, trans. H. T. Clarke and L. Silberstein (Oxford: Clarendon Press, 1922). Reprinted in *The World of the Atom*, ed. H. A. Boorse and L. Motz (New York: Basic Books, 1966), 496–500.

Chapter 13. Other Factors in Theory Evaluation

1. Thomas Kuhn, *The Essential Tension* (Chicago: University of Chicago Press, 1977), 329.

2. Francis Bacon, *Novum Organum,* book 1, aphorism 68.

3. Lord Kelvin (William Thomson), *Lectures on Molecular Dynamics and the Wave Theory of Light* (Baltimore: Johns Hopkins, 1884), 131–32.

4. Pierre Duhem, *The Aim and Structure of Physical Theory* (New York: Atheneum, 1962), 84–85.

5. The evidence for the impact theory is discussed in William Glen, "What the Debates Are About," in *The Mass-Extinction Debates,* ed. by William Glen (Stanford, Calif.: Stanford University Press, 1994) 9–16.

6. William Glen, "How Science Works in the Mass-Extinction Debates," in *The Mass-Extinction Debates,* ed. William Glen, 88–91.

7. Gerald Holton, "Thematic Presuppositions and the Direction of Scientific Advance," in *Scientific Explanation,* ed. A. F. Heath (Oxford: Clarendon Press, 1981), 17–25; *Thematic Origins of Scientific Thought,* rev. ed. (Cambridge, Mass.: Harvard University Press, 1988), 10–68; *The Scientific Imagination* (Cambridge: Cambridge University Press, 1978), 6–22.

8. Galileo, *The Assayer,* in *The Controversy on the Comets of 1618,* ed. S. Drake and C. D. O'Malley (Philadelphia: University of Pennsylvania Press, 1960), 183–84.

9. Albert Einstein, "Reply to Criticisms," in *Albert Einstein; Philosopher-Scientist,* ed. P. A. Schilpp (New York: Tudor, 1949), 667–73.

10. Max Born, *Atomic Physics* (New York: Hafner, 1946), 87.

11. Neils Bohr, *Atomic Physics and Human Knowledge* (New York: John Wiley & Sons, 1958), 46.

12. Ibid., 72.

13. Ibid., 88.

14. Holton, "Thematic Presuppositions and the Direction of Scientific Advance," reprinted in *The Advancement of Science, and Its Burdens* (Cambridge: Cambridge University Press, 1986), 19–27.

15. Roger Boscovich, *A Theory of Natural Philosophy,* 1763 (Cambridge, Mass.: MIT Press, 1966), 22.

16. Einstein, *Out of My Later Years* (New York: Philosophical Library, 1950), 63–64.

17. Holton, *The Scientific Imagination,* 7–8.

18. Holton, *The Advancement of Science, and Its Burdens,* 70–72; *Thematic Origins of Scientific Thought,* 252–54.

19. Paul Feyerabend, *Against Method* (London: NLB, 1975), 23–24.

20. John Worrall, "Fix It and Be Damned: A Reply to Laudan," *Brit. J. Phil. Sci.* 40 (1989), 383.

21. Isaac Newton, *Opticks,* 1730 (New York: Dover, 1952), 154–58.

22. Worrall, "The Value of a Fixed Methodology," *Brit. J. Phil. Sci.* 39 (1988), 274.

23. Worrall, "Fix It and Be Damned," 386.

24. Ibid., 380.

25. Lewis Carroll, "What the Tortoise Said to Achilles," *Mind 14* (1895), 278–80; 27.

Conclusion

1. Airy's position is outlined by Morton Grosser, *The Discovery of Neptune* (Cambridge, Mass.: Harvard University Press, 1962), 48–49, and by Alan Musgrave in the excerpt "Musgrave on 'Hard Cores' and the Newtonian Research Program" in chapter 9 of the present volume.

Index

normative judgments, 2
Novum organum (Bacon), 43, 167
nuclear model of the atom, 21–23

observation report, 29–30, 64
octaves, law of, 66
Oersted, H. C., 98–100
operational requirement, 186–87, 192
Opticks (Newton), 42
Origin of Humankind (Leakey), 38
orthogenesis, 34, 36
"out of Africa" hypothesis, 38–40
ovism, 13–14

paradigm, 93–94, 101
parity, 95–97, 182
Pascal, Blaise, 44–46
Pasteur, Louis, 95
Pauli, Wolfgang, 77–78
Peierls, R. E., 125–26, 135–37
pendulum, 70–71
Perier, Florin, 45–46
periodic law of chemical elements,
   66–69, 77, 103
Perrin, Jean, 159–62, 166
perturbations of planets' orbits, 60, 87
phenomena, kinds of, 88–89, 153–54,
   191
phlogiston, 20–21, 109–13, 149–50,
   153, 159, 182
photoelectric effect, 132, 146–69
photons, 146
pictorial level, 179–81
Planck, Max, 131–32, 146–47, 154,
   162–66
plate tectonics theory, 114
Plato, 7
plenism, 183
"plum pudding" atom, 21–22
Poisson, Simeon, 142, 151
polarization, 84, 87, 115, 168
polydactyly, 14–15
Popper, Karl, 63–65, 74–77, 105, 123,
   186, 189
positive heuristic, 104, 106–9, 119
precession of equinoxes, 85

prediction, 16, 167, 175, 188; and falsification, 8, 27, 28; Mendeleyev on,
   69; novel, 110, 112, 125, 135–51,
   153–54
predictivist thesis, 137–38, 141–53
preformationism, 13–14
Priestley, Joseph, 18–21, 23, 111–12,
   149–50
prism experiments, 40–41
progress, 83–89, 191
protowings, 47–50
Prout, William, 75, 116, 123
Prout's hypothesis, 75, 116
Ptolemy, Claudius, 31, 85, 114, 126,
   149, 159
punctuated equilibria, theory of, 174

quantum theory: and Bohr's theory of
   the hydrogen atom, 126–27, 132;
   and Einstein, 177–79; and formalism, 130; and Planck's hypothesis,
   146–48, 154, 162–65, 182
quarks, 123
Quine, Willard van Orman, 30–32, 189

radioactive decay, 94–95, 160
rational reconstruction (Lakatos), 119,
   123
Rayleigh, Lord (J. W. Strutt), 132
redefinition, 70, 94
refraction, 41–42, 71, 87, 155, 168
revolution, 90
Rey, Jean, 111–12
Ruhe, Jacob, 13–14
Rumford, Count (Benjamin Thomp-
   son), 9–13, 23
Rutherford, Ernest, 21–23

sap, circulation of, 8–9
Scheele, Carl, 111
Schrödinger, Erwin, 130, 179, 181, 183
scientific research programs (Lakatos),
   102–16, 123–24
seafloor spreading, 143–45
"sea of air" hypothesis, 45–46
Shapere, Dudley, 117